Please don't get lost.
This has nothing whatever to do
with the archaic flat-Earth concept
of the Ptolemaic kings . . .

Physical
Worlds
Continuity
Beyond
of
the
the
Poles
Universe

F. AMADEO GIANNINI

HEATHEN EDITIONS
THEIR BOOKS. OUR WAY.

Published in the good ole United States of America
by Heathen Editions, an imprint of
Heathen Creative
P.O. Box 588
Point Pleasant, WV 25550-0588

Heathen Editions are available at quantity discounts.
Bear witness to the yackety-yak and tomfoolery at:

heatheneditions.com

Social? Tag us! @heatheneditions
Photo? Tag it! #heathenedition

Caution: This book may alter your mind.

First published July 6, 1959
Heathen Edition published September 5, 2025

Paperback ISBN: 978-1-948316-47-7
Hardcover ISBN: 978-1-963228-47-2

Book and cover design by Sheridan Cleland
Set in 10pt Droid Serif
Titles in Draft Natural

FIRST HEATHEN EDITION

Heathenry
Thoughts on the Text

If, dear reader, you've never fallen down some rabbit holes of the "alternative science" or "conspiracy" variety, then the hypothesis presented in this book may be difficult to digest.

If, however, you are familiar with flat earths, mud floods, star forts, phoenix cycles, space bloopers, and strange lights in the sky, then you're about to find yourself comfortably right at home.

Either way, we must warn you: *this is still a difficult book* because Giannini doesn't make understanding his concept simple and he piles on complexity with maddening repetition, which is why we have included, as appendices, ten relevant articles that we believe will vastly enhance your reading and understanding of his concept.

However, before we explain why we have included those articles, let's establish some basics.

First, who was F. Amadeo Giannini?

Which is a great question because after all of our research, we Heathens are still not sure who he was, exactly.

Born Francis Armadeo Johnnene on October 3, 1898, in Cambridge, Massachusetts; his father, Giter Johnnene, was a violinist from Italy and his mother (Giter's second wife), Mary Elizabeth McClafferty, was Canadian.

He had an older sister and two younger brothers, and

then his father died in 1904 when Francis was only 5 years old.

He had a half-brother and half-sister from his father's first marriage, and when his mother remarried in 1906 there soon followed two additional half-sisters and two half-brothers. He was educated — his writing broadcasts it — but where we don't know. Some reports cite Yale, others Harvard, also Oxford.

It's hard to imagine that he had any time for an ivy league education considering he married his first wife, Ethel Marie Hanson, in 1916 at the age of 17, with their first son arriving at 18, and the second at 19, but we only know what we know.

He and Ethel split in 1923, then he married his second wife, Emily Adeline Nash (née Foss), in 1927.

Here, we get our first glimpse at how Francis supported the family as a marriage announcement identifies him as "a salesman, of Boston, Mass."[1]

He makes his next newspaper appearance in 1928 sporting a new name — Francis Amadeo Giannini [seen at right][2] — likely to capitalize on the similar-sounding name of another Italo-American who was making financial headlines at the time: banking legend Amadeo Pietro Giannini had founded the Bank of Italy, which was then merging with and becoming Bank of America.

Remember: *a salesman.*

Also in 1928, there appeared the first of the ten articles that we have included — "Fantastic Plan for a Trip to Mars" — which was syndicated in several papers during November of that year, wherein Giannini's concept is first presented for public consumption. We have situated it as an introduction because we find that it's a good primer for understanding

---

[1] Mrs. Emily Nash Secures License — Brattleboro Telegraph Operator to Wed Francis A. Johnnene, Boston Salesman — Goes to New York. (1927, September 13). *The Brattleboro Daily Reformer* 15(167). A1.

[2] Says Earth is No Globe. (1928, August 3). *Oakland Tribune* 109(34). A35.

Says Earth Is No Globe

FRANCIS A. GIANNINI (left), youthful Boston philosopher, and CAPTAIN GEORGE WILKINS, polar aviator, at the Breakfast club in Los Angeles. Giannini believes the world stretches on through space past the poles. Wilkins will have a chance to test the theory when he takes his projected flight to the south pole.

—*Associated Press photo.*

the book to come as the entire concept is condensed and summarized for the layman.

What's important to note here is that Giannini consistently points to the discoveries of the arctic and antarctic expeditions of George Wilkins and Richard Byrd as evidence that his hypothesis can be proved.

The first of those being Wilkins' December 1928 Antarctica expedition, which today's media glosses over as a simple-yet-history-making flight over the continent.

Those of you who are currently comfortably at home (as mentioned at the start) will surely find it interesting that Wilkins' base of operations for that expedition was located on Deception Island.

Then, the 1930s arrived and Giannini's life, as told via a series of newspaper articles, seemed to become as difficult as understanding his concept.

If you read our blurb for this book, then you've probably been wondering about that mention of Giannini being a "four-time convicted burglar." Well, buckle up, here goes . . .

On April 3, 1930, newspapers nationwide reported Giannini had been found guilty of burglary in Michigan, where he apparently went to "get drunk and forget my troubles" and "registered at a hotel under an assumed name, being afraid I would get in trouble."[3]

Right, so — forgetting your troubles by *getting into trouble* — this may surely get *interesting*.

As usual, per our research, the newspaper reports are conflicting, but we'll opt for the details as reported in *The Detroit Free Press* since they were closest in proximity to the happening:

> "According to the prosecution, Giannini prowled through apartments wearing a dressing gown and,

[3] Relativity Expert Found Guilty of Breaking Entering. (1930, April 3). *Indiana Evening Gazette* 26(187). A6.

Finds Yale Graduate Guilty of Burglary

Kilpatrick Refers Case of 'Philosopher and Writer' to Psychopathic Ward.

Graduate of Yale, lecturer philosopher, scientist and magazine writer, Francis A. Giannini, 35 years old, was found guilty in recorder's court Wednesday of having been the "sleep walking burglar" who robbed several Pallister and Seward avenue apartments. Judge Arthur W. Kilpatrick referred his case to the psychopathic and probation departments.

According to the prosecution, Giannini prowled through apartments wearing a dressing gown and, if he aroused the occupants, he started, rubbed his eyes and generally acted as if just awakened. Then he would apologize for the intrusion, explaining that he was a somnambulist and "guessed he had wandered out of his own apartment," while asleep.

Although convicted of having broken into and entered the apartment of Nina Cassidy, 148 Pallister avenue, several witnesses testified that he had broken into other apartments. When he was arrested, two weeks ago, police said they found on his person a $20 bill identified by a tenant at 50 Seward avenue as one stolen from his apartment. Giannini lived at 50 Seward avenue.

The defendant testified that he "did not remember" anything of the alleged entry into the Pallister avenue apartment. He said that on March 12, date of the offense, he had "some wine and some girls," in his apartment and that he could not remember leaving the apartment.

Among Giannini's effects police found numerous magazine articles he had written and many newspaper clippings showing that he had been interviewed as a scientist and philosopher. When he was arrested he claimed relationship with the Giannini family prominent in Pacific coast banking circles.

RELATIVITY EXPERT FOUND GUILTY OF BREAKING ENTERING

DETROIT, April 2.—INS—Francis A. Giannini, Yale graduate, and lecturer on relativity, astronomy, and Polar exploration, and an author of scientific books, was found guilty of breaking and entering by a jury in Judge Arthur W. Kilpatrick's court yesterday. Giannini claims he is a cousin of A. P. Giannini, one of the country's most powerful bankers and head of the trans-American bank.

He said he came here from Cleveland March 12 with only $75 in his pockets.

"I intended to get drunk and forget my troubles," he told the jury, "so I registered at a hotel under an assumed name, being afraid I would get in trouble."

Police said he broke into a tea room, and several rooms in a hotel, stealing small amounts of money. He was arrested immediately.

"I may have done those things," Giannini said. "I was too drunk to remember."

FRANCIS A. GIANNINI.

Lecturer Tried For Burglary, Found Guilty

DETROIT, Mich., April 2.—(UP) Charged with breaking and entering, Francis A. Giannini, university lecturer on Polar exploration and astronomer, was found guilty by a circuit court jury here today. Giannini, who said he was a cousin of A. P. Giannini, prominent San Francisco banker, was not sentenced. Judge A. W. Kilpatrick referred the case to the probation and psychopathic divisions for investigation.

BURGLAR

Burglar Francis Giannini, who claims to have written "The World Beyond the Poles" and "The Continuity of the Universe," two books on philosophy, is soon to be released from Michigan state prison. He was jailed for ransacking rooms of a Detroit hotel. He says he has lectured against the Einstein theory.

RECOMMENDS RELEASE

JACKSON ,Mich., June 10.—(AP) —Prison officials were advised today that Parole Commissioner Howard J. Waples had recommended the release of Francis Giannini, who has written two books on the Einstein theory since he was sentenced to Michigan state prison for breaking and entering a year ago. Information received here was that Giannini, a college graduate, will be permitted to go to his mother's home in Boston, where he hopes to have his books published. Giannini was sentenced from Detroit in March, 1930, to serve from 18 months to 15 years for thefts committed in several hotels.

CRIMINAL COURTS
New York County

Francis A. Giannini, 32, of 230 W. 101st St., to an indeterminate term, in the penitentiary, not to exceed three years; by Justices Daniel A. Direnzo, Max Salomon and Charles P. Caldwell; for unlawful entry.

Not Circulating

New York, May 6.—(UP)—One of the town's colorful characters was removed from circulation today when Francis Giannini, 32, was sentenced to the penitentiary for a term not exceeding three years after he had failed satisfactorily to explain his presence in somebody else's suite at the Hotel Plaza.

In 1928, Giannini made the Sunday supplements with his theory of inter-planetary bridges and his advice to Commander Richard E. Byrd to look for a terminal of a highway to Mars at the South Pole. Giannini explained his presence in the Hotel Plaza by saying that he had mistaken it for the Hotel St. Moritz.

Author Paroled On Drunk Theft Charge

(By The Associated Press)

Lansing, June 10—A parole has been granted Francis Giannini, Harvard graduate and author of books on philosophy, effective July 31, when his minimum term expires, it was announced Tuesday by Harold Waples, parole commissioner. Giannini was sentenced from Detroit more than a year ago to serve 1 1-2 to 15 years for breaking and entering. It was claimed he wandered from room to room in a hotel, picking up small sums of money while intoxicated.

BOSTON WRITER WANDERS IN HOTEL

Giannini Tells N Y Police He Was Jagged

NEW YORK, Jan 7 (A. P.)—Francis A. Giannini, who said he was a writer and scientific research worker and gave 16 Beacon st, Boston, as his address, said in the police lineup today he was intoxicated when he was found wandering on an upper floor of the Hotel Plaza last night.

He said he thought he was in the St Moritz Hotel, which is near the Plaza on West 59th st, facing Central Park, and that he was looking for a friend. Police said, however, that no one by the name he gave for his friend was registered at either hotel.

He admitted to police today he had been arrested under similar circumstances in a Detroit hotel, but said the case was dismissed.

No trace of a Francis A. Giannini could be found today in any of the Boston city directories. The address, 16 Beacon st, which Giannini is said to have given New York police, is the former headquarters of the Unitarian Laymen's League and the Unitarian Foundation, Inc, but is now unoccupied.

if he aroused the occupants, he started, rubbed his eyes and generally acted as if just awakened. Then he would apologize for the intrusion, explaining that he was a somnambulist[4] and 'guessed he had wandered out of his own apartment,' while asleep."[5]

Remember: *a salesman.*

But the judge wasn't buying it and sentenced him to one and a half to 15 years for breaking and entering.

14 months later, "Harvard graduate" Giannini was granted parole.[6]

Then, a mere six months later, he was back in the papers for Drunken Burglary: Round 2, this time in New York,[7] and subsequently sentenced to "a term not exceeding three years."[8]

How long he actually served we don't know, but two and a half years later he makes an appearance in a bizarre gossip column-like article entitled "Play-Writing Wife Sued in Divorce Drama" in the New York *Daily News* for December 26, 1934.

Gone is the second wife because here appears a third wife, costume designer Lucille Giannini, who along with Francis and a former real estate broker were routine visitors to the apartment of the aforementioned play-writing wife, who had apparently taken a liking to the broker, which caused the ensuing divorce drama for which the Gianninis found themselves as chief witnesses when the play-writer husband filed suit against his wife.

[4] One who sleepwalks.

[5] Finds Yale Graduate Guilty of Burglary. (1930, April 3). *The Detroit Free Press* 99(334). A10.

[6] Author Paroled on Drunk Theft Charge. (1931, June 10). *The Port Huron Times Herald*. A4.

[7] Boston Writer Wanders in Hotel. (1932, January 7). *The Boston Evening Globe* 121(7). A6.

[8] Not Circulating. (1932, May 7). *The News and Observer* 134(128). A2.

ch. ends p. xxv

Since everyone involved seemed hellbent on throwing each other straight under the bus, the play-writing wife stated in her affidavit: "Giannini was a man of many aliases, who posed as a Doctor of Philosophy from Harvard and a nephew of A. P. Giannini, the West Coast banker." With her attorney producing evidence which "purported to show that Giannini was an ex-convict and a fugitive from justice."[9]

One year later, Giannini's '30s problems climaxed when he found himself arraigned on a third charge of burglary in a Chicago courtroom, wherein he described himself as a "cosmic philosopher," whereat the judge told him "to philosophize over a $10,000 bond pending his trial."[10]

Cosmic, indeed, for we Heathens can feel the heat from that burn across time and space . . .

Then, all Giannini reports go quiet for 12 years, until he pops up in the next article that we have included, "On the Line with Considine," which was syndicated nationally in January 1947. It's a short but curious article because Bob Considine approaches his subject with a healthy dose of skepticism and we learn that Giannini had been, allegedly, "an Antarctic chartmaker" for George Wilkins, although we have been unable to locate any further information to verify that claim.

Additionally, Giannini states the reason for his "sudden" reappearance is a direct result of the first photos taken from space by the V-2 No. 13 rocket launched on October 24, 1946, in White Sands, New Mexico: "I now have absolute proof that the universe is one body," he said; adding, "I have been silent for 20 years because I lacked visual proof of my theory."

Essentially, and without spoiling too much of the book ahead, Giannini claimed that those images displayed the

[9] Play-Writing Wife Sued in Divorce Drama. (1934, December 26). *Daily News* 16(157). A35.

[10] Chance to Philosophize. (1935, August 15). *The Pittsburgh Press* 52(53). A10.

"globular illusion" created by a round, globular camera lens, itself fashioned after the round, globular human eye, and therefore subject to its same errors of perception.

The problem that we've found with this argument is the single photo from that 1946 launch which Giannini includes in the preface of this book to illustrate the "proof":

Now, you may be thinking that the above image lacks globularity of any kind since one can draw, diagonally, a straight line across the pictured horizon, but that is to misunderstand Giannini's hypothesis — which is that if one travels far/high enough into the stratosphere, then the area of earth directly below the observer will always take on a globular appearance similar to the below stitched-together panorama, taken by another V-2 rocket launched on July 26, 1948:

ch. ends p. xxv

His notion being that the black area in the photos that we assume is the void of space is simply the areas of earth beyond the limit of our — and, in kind, the camera lens' — perception.

Following?

If not, here's how Giannini describes it in Chapter 8:

One must never lose sight of the fact that there exists no observing instrument that was not patterned after the human lens. The human lens is great and magnificent; but it is subject to many errors. Therefore, it must be held in mind that every lens holds the same elementary error as the optic lens. It demonstrates gross misunderstanding to claim that though the human lens is subject to error, the photographic lens overcomes the inherent error. It does no such thing. If it did, there would not be curves developed by the photographing lens.[11]

One way that we've found to think about this portion of his hypothesis, especially given today's camera technology, is to envision the perception of a 360° camera, which renders its extreme view into what is perceived as a complete circle or sphere — *a globular illusion* — with left to right (and vice versa) straight-line movements near the camera appearing to advance in clockwise or counter-clockwise fashion on camera.

Another way to think about it is the "globular illusion" created by the fisheye lens of most GoPro cameras, whose software has a setting which allows for the automatic correction of its globular distortion, returning its recorded arched horizons to the flat, straight lines we actually observe.[12]

Say what you will about the rest of Giannini's hypothesis

---

[11] pp. 153-154
[12] Insert NASA joke here.

(and/or his criminal tendencies), but his belief that what we perceive with our round, globular eyes including what is perceived by all round, globular camera lenses is actually distorted and, therefore, inherently flawed has brute-forced us into reevaluating our entire perception of reality.

Or "reality."

Like they say on the street: Shit's deep, yo.

If this still isn't making sense, the next article that we've included may help: "Can We Walk to Mars, Maybe Jupiter and the Other Planets and Stars" was syndicated in various forms and with varying titles in 1947,[13] soon after the "On the Line" article. It's probably the best condensation of Giannini's hypothesis as it's more in–depth than his earlier 1928 article and, again, written (by Giannini, no less) for laymen comprehension. We've situated it as the first appendix because we believe it's an excellent recap and summation of the primary text, and a good setup for the appendices that follow — the next of those being "Giannini's Theory and Byrd Findings" as featured in *The Rock Island Argus* for April 4, 1947. We include it because we believe the managing editor's approach to the criticisms received by *The Argus* for publishing Giannini's hypothesis is both sound and level-headed. It's also proof that Giannini was never without his critics (and still isn't).

After this round of newspaper action in 1947, all Giannini news goes quiet again until he pops up in early 1955 for his fourth and (as near as we can tell) final court appearance, this time for a third degree burglary charge, which netted him "six months to three years" in the Utah State Prison.[14]

What's interesting about this episode is that, in contrast with his three previous sentences, Giannini actively fought

[13] An ad for the feature, as seen in the *Editor & Publisher* for March 8, 1947, can be found on the next page.

[14] Suspect Files Innocent Plea on Deadly Weapon Count. (1955, January 16). *The Salt Lake Tribune* 170(94). B4.

ch. ends p. xxv

Are there
WORLDS BEYOND THE POLES?

Can we reach Mars, Venus, Saturn and the other planets?

Did Adm. Byrd step off this earth?

.

FRANCIS A. GIANNINI,

Scientist and Philosopher, Says We Can Get from Planet to Planet by Moving on the Same Level!

Here is the first man in 500 years to give us a greater picture of the universe!

His theory of the physical continuity of the universe is one of the most amazing stories ever written!

.

3 daily articles—or combine as Sunday magazine feature. Unusual illustrations.

.

Like to see copy? Release date is March 30. Phone or wire

FRANCIS Armadea Johnnene was committed to prison for from six months to three years. He had been found guilty of burglary in the third degree.

SUITS FILED
Third District Court

Francis Armadeo Johnnene vs. Marcell Graham, warden, Utah State Prison. Plaintiff seeks writ of habeas corpus, for alleged illegal and unlawful restraint.

Inmate at Prison Denied Petition

UTAH STATE PRISON — Francis A. Johnnene, an inmate at the Utah State Prison, Thursday was denied a petition for writ of habeas corpus by Judge Sherman A. Christenson in U.S. District Court.

Johnnene was sentenced to the prison on a third degree burglary conviction Jan. 5.

In court, the prisoner claimed an attorney who represented him at the jury trial failed to present all the facts. He also said witnesses presented perjured testimony.

Judge Christenson, in denying the writ, said the prisoner should exhaust all remedies within the state low before bringing the case to the district court.

Understanding Unit Lecture Thursday

F. Amadeo Giannini, scientist, philospher and author, will deliver a lecture entitled, Worlds Beyond the Poles, to members of the Understanding Unit No. 18, at 8 p.m., Thursday, at 839 Third st this ruling,[15] arguing that "an attorney who represented him at the jury trial failed to present all the facts" and "witnesses presented perjured testimony."[16] So strongly did Giannini feel he was wronged that he sought a writ of *habeas corpus*[17] on four separate occasions, only to have it dismissed all four times.

We believe there are two ways you can look at this particular episode: the first by applying that old adage about a leopard and its spots, the second by applying a bit of conspiracy-think with the presumption that maybe someone or *someones* wanted to shut him up — at least momentarily.

Perhaps both are true?

Either way, it seems the entire episode emboldened and fortified his resolve because his next appearance on the world stage is with the 1959 self-publication of his one and only book: *Worlds Beyond the Poles*.

That it was self-published is confirmed by Giannini himself at "an expenditure of $3,000" (approximately $33,000 in 2025 dollars) in one of the six final articles that we've included, all from Ray Palmer's[18] *Flying Saucers* magazine.[19]

If you're not familiar with Admiral Byrd's arctic and antarctic expeditions, then the first article — "Saucers from Earth!"[20] — is a great overview by Palmer as it recounts many of the initial "Hold up: *Say what?*" details from the very

[15] News of Record: Suits Filed. (1955, September 2). *The Salt Lake Tribune* 171(141). B3.

[16] Inmate at Prison Denied Petition. (1955, June 10). *Desert News* 343(138). A12.

[17] The civil right one may seek to be brought before a court to obtain immediate release from an unconstitutional confinement or illegal imprisonment.

[18] Raymond Arthur Palmer (1910–1977) was a prolific American author and editor, best known for the magazines and books he edited and published, some of which include *Amazing Stories, Fantastic Adventures, Fate Magazine, Imagination, Mystic, Other Worlds, Space World,* and *Flying Saucers.*

[19] *Flying Saucers* was published with varying titles by Palmer Publications, Inc. in Evanston, Illinois, from 1957 to 1976.

[20] Palmer, R.A. (1959, December). Saucers From Earth! *Flying Saucers,* pp. 8–21.

ch. ends p. xxv

first radio broadcasts and news reports of those expeditions that seem to have now been stricken from the official record and/or lost to time.

If you're still comfortably at home, then what you may soon realize is that those "lost" details are comparable to today's many and multiplying Mandela effects because as Palmer notes: "Millions heard the radio broadcast description of that flight, which was also published in the newspapers."

That article, which was featured in the December 1959 (FS-13)[21] issue, kicked off a Giannini saga of sorts for *Flying Saucers* magazine, which played out via Palmer's editorials over the course of all of its 1960 issues — February (FS-14), June (FS-15), August (FS-16), and November (FS-17) — all of which we have included.

It's a fascinating narrative when strung together, back-to-back, because it seems the then-readers of *Flying Saucers* really took Palmer to task for his "Saucers from Earth!" article, many claiming it all a bunch of poppycock, which led to some mild name-calling, a bit of backpedaling, and Palmer pointing his finger squarely at Giannini for all the polar hoopla, while at the same time Palmer seemed to dig his heels in and started asking some hard questions about the "official" arctic and antarctic polar narratives.

And Palmer's problems arising from the publishing of "Saucers from Earth!" didn't simply end with the reader uproar. What he relays and slowly brings into focus over the course of his four 1960 editorials is that there appeared to be a coordinated effort across multiple fronts by actors unknown wholly determined to disrupt the routine publishing of *Flying Saucers*.

All of which culminates in the final article, taken from the February 1961 (FS-18) issue, that closes out our edition

---

[21] Each issue was identified by this code, making the December 1959 issue *Flying Saucers* #13.

— "Byrd Did Make North Pole Flight in Feb. 1947!"[22] — which constitutes a series of letters between Giannini and one Richard Ogden, a reader of *Flying Saucers*, with relevant commentary by Palmer. The primary contention of the article being an alleged North Pole flight by Admiral Byrd in February 1947. Of course, the "official record" states there was no such flight, but Giannini swears there most certainly was by citing anecdotal evidence and ancillary *New York Times* articles, which forced Ogden to dig for and provide further evidence, which then led Palmer to exclaim: "Hold up: *Say what?*"

We believe the *Flying Saucers* articles and editorials are essential because, again, Giannini routinely points to the initial revelations of the arctic and antarctic expeditions by Admiral Byrd as evidence that his hypothesis can be proved, and as Dr. Raymond Bernard[23] notes in his book *The Hollow Earth*: "The true significance of Admiral Byrd's great discoveries was hushed up soon after he sent his radio report from his plane, and was not given the attention it deserves until Giannini and Palmer publicized the matter."[24] And while some of the *Flying Saucers* narrative may, indeed, be poppycock, what, then, of the coordinated effort by actors unknown? And why all that funny business *immediately after* the "Saucers from Earth!" article was published? Correlation doesn't prove causation, as they say, but if one finds themselves suddenly getting stung six, seven, a dozen times, then the odds are fairly likely that one has stepped squarely into the middle of a yellow jacket's nest — *and they're pissed.*

Based on our research, we've found one thing to be

---

[22] Byrd Did Make North Pole Flight in Feb. 1947! (1961, February) *Flying Saucers*, pp. 4–11.

[23] Walter Isidor Siegmeister (1903–1965), later known as Raymond W. Bernard, was an American alternative health advocate and esoteric writer, who formed part of the alternative reality subculture and is credited with the merger of the Hollow Earth theory and religious beliefs about UFOs.

[24] Bernard, R. (1969). Foreword. *The Hollow Earth* (p. 25). University Books, Inc.

ch. ends p. xxv

absolutely certain: no matter how you ultimately feel about Giannini or his hypothesis, the deeper down the Admiral Byrd arctic/antarctic rabbit hole you go, the stranger it gets — and for those of you who are *still* comfortably at home, you'll surely discern the clues like high beams illuminating reflective road signs on a cold clear night . . . *Following?*

Now, moving on, one thing we should note outright is that, beginning in Chapter 1 and throughout the rest of the book (excluding the appendices), Giannini routinely refers to himself both in the first person and in the third person as "pilgrim," "seeker," or "dreamer," which, if you're not prepared for it, can provoke some mild confusion.

As for the text, we've modernized a few hyphened words so they're easier on your eyes (co-operation has become cooperation, and so on), and to reduce even more confusion than we think is necessary we have altered certain spellings of names and places so that they appear consistently the same throughout the entire book *and* the appendices.

Typically, we're proponents of *less* hyphens, however, we have altered one recurring phrase in this book to *always* feature a hyphen and that is "sky-light." There was no real consistency with Giannini's hyphen usage when it came to that phrase in his original work, but he does consistently mean the same thing — the light of the sky, whether terrestrial or celestial — and since his usage and application of that phrase is unique to this narrative, we felt including the hyphen is a great reminder, as a reader, that this particular "sky-light" is Giannini's sky-light.

Additionally, Giannini's original work featured five hand-drawn illustrations (**Figures 1**, **2**, **3**, **4**, and **6**) that we, at first, attempted to restore to their original glory, but, then, ultimately decided to recreate from scratch so as to bring them more in-line with the overall design aesthetic of our edition.

You may have also noticed the many newspaper facsimiles collected collage-like that we have already included, which

Byrd Drops United Nations Flag At South Pole; No Explanation is Given

By ALTON L. BLAKESLEE
Associated Press Correspondent
Representing the Combined
American Press

ABOARD THE U. S. S. MT. OLYMPUS IN THE ANTARCTIC—(AP)—Rear Amiral Richard E. Byrd flew over the South Pole today and dropped the flag of the United Nations at the Pole, dispatches from Little America announced.

In flying over the Pole, Byrd duplicated his feat of Nov. 29, 1929. The dispatches did not state how far his plane had flown into areas never before beheld by human eyes.

Byrd returned to Little America at 11:45 a. m.—which is 5:45 p. m., Eastern Standard time—after a flight of nearly 13 hours.

His plane took off accompanied by a second craft and returned with it, but it was not immediately announced whether both had flown over the pole.

Byrd sent a personal message to Admiral Chester Nimitz, U. S. chief of naval operations, written while he was circling over the pole.

The temperature at the pole was reported to be 40 degrees below zero at an altitude of 12,000 feet.

The significance of Byrd's dropping the United Nations flag at the South Pole was not explained. Radio communications with the Little America base and this headquarters ship was difficult most of Sunday due to atmospheric conditions.

When the admiral arrived at Little America Jan. 30 on a hop from the carrier Philippine Sea, he told correspondents:

"I'd like to see the land beyond the pole. It is the most inaccessible area on the face of the earth. I would like for somebody on this expedition to get beyond the pole into unknown territory. They might find anything there. There might be a lot of volcanic mountains. There might be some of the highest mountains or the highest plateaus in the world. That area beyond the pole from Little America is the center of the unknown. You might see something there you had never seen before."

NEW CONQUESTS

The discoveries recently made by Admiral Byrd and his flyers in the frozen south-polar continent are interesting, especially the remarkable inland "oasis of muddy pea green lakes dotted with tall, dark brown mounds of apparently bare earth," reported the other day. There will be new opportunities for years in that formerly unknown region, and doubtless many volunteers to explore and exploit it.

Rear Admiral Richard E. Byrd, America's foremost contemporary Polar explorer, believes that the United States must take every precaution to guard against future attacks from the "bottom" and "top" of the world if another war should come.

Byrd Receive Masonic Award

NEW YORK, May 7—(AP)—Rear Admiral Richard Byrd has been presented with the distinguished achievement medal of the New York State Masonic Grand Lodge. The medal was awarded to Byrd last night in recognition of his high achievement as an outstanding explorer, scientist, author and naval officer.

In receiving the award, Byrd called for a spirit of brotherhood among all the peoples of the world. The Admiral added that the United Nations organization must be made to work or, as he put it, "Civilization will go down in darkness."

we also feature throughout the rest of the book, and which we believe adds authenticity to many of Giannini's claims, however absurd you may find them, and sometimes supplements or expands upon the primary text with additional, interesting tidbits and details.

We've also included cover facsimiles of each issue of *Flying Saucers* from which we have extracted the articles and editorials for the appendices.

Finally, the true bulk of work lies in the 330+ footnotes that we have appended throughout the entire text to identify our source materials where necessary, and to provide clarity, context, and commentary, especially as certain portions of this edition becomes a literal *Who's Who* of science and polar exploration.

In all, what you now hold in your hands represents over two years of our scholarly-like investigation — through archives, contradictions, and curiosities — and is, without any doubt whatsoever, our most scrutinized and researched **Heathen Edition** to date — all 400+ pages of it.

We hope our work speaks for itself . . .

Godspeed!

NEW SHAPE FOR EARTH?

By Wilbur B. Mueller.

Francis A. Giannini, who is described as a philosopher and scientist, is said to be the first man in centuries to have proposed a radical new theory relating to the earth and planets. According to Giannini the earth is physically connected not only with the planets such as Mars and Venus, but also with the stars, the nebulae and the galaxies.

We call attention to the Giannini theory here, not to recommend it or to argue against it, but merely to give a shock to those who have looked at planets and stars through telescopes. Looking at a planet through a telescope, you apparently see a globular object far off toward the stars. If you have a telescope powerful enough, you can see the rotation of Mars on its axis, and see its snow-capped poles. Mars appears to be a globe disconnected from the remainder of the solar system. But, according to Giannini it's all an optical illusion. Actually, he implies, Mars can be reached without leaving the earth.

I once became worked up about attention paid to the Einstein theory of relativity, and figured that a person couldn't fairly be called even an amateur astronomer unless he had read it. So after a great many hours of intense mental agony it became apparent that if I had to understand that theory in order to be an amateur star-gazer, I'd pass up the reputation. And anyway, Professor Einstein revises his theory now and then just to keep amateurs on their toes. In the Einstein theory I found statements which I thought could be disproved by the application of mere common sense—but it is a notorious fact that professional astronomers pay him the greatest respect and consideration. If the Giannini theory is to be disproved, let the professionals do it.

In a southern city there used to be an energetic, self-assertive man who declared the earth was flat. It was in vain to tell him that a ship moving away from a harbor does not disappear in perspective, but seems to sink into the sea, thereby proving the roundness of the earth. He said it was the waves. He admitted that the moon was round because he could see it. I once tried arguing with him, using all the school book statements, but he had one devastating argument. He would ask whether I had ever been around the earth, and say that since I hadn't, I ought to shut up. Giannini has an argument, too.

He indicates that if you were to take an airplane and fly "over the pole," as you'd put it, you would first be going north, and later south, thereby proving that you had changed directions.

It would be comforting to know that Giannini was right. If, as he says, all the planets, etc., are physically connected with the earth by some sort of polar attachments, all we need do is keep going straight north or south and finally we'd reach another planet. He says they are provided with light and air like ours. Maybe the lack of geographical frontiers can be overcome and the crowded nations of the earth can be relieved of their pressure.

It is interesting to note that Giannini evolved his theory years ago, and it got attention only because of the discovery by the Byrd expedition of certain "oases" near the south pole, where there was no snow and where some strange animals were believed to have been sighted.

It is also interesting to note that both Einstein and Giannini could have been committed to the hoosegow some centuries ago for being so bold as to challenge popular opinion.

New Universe Continuity Theory Is Expounded.

By NADIA LAVROVA.

FATHER JEROME S. RICARD has lent his attention to the young Italian-American who is trying to prove that this earth, far from being a globe suspended in space, extends in an endless trunk through the universe, linking planets to one another, to the sun, and perhaps to other suns.

"If you succeed in proving this, it will be the most absolute continuity imaginable," said the bright-eyed scientist and philosopher of the University of Santa Clara, after listening to Francis Amedeo Giannini develop his astounding theory. The latter claims that his conception is an improvement on that of Copernicus.

"The great fact is that there is no end to this spot, which we dwell upon, known as the 'earth' or the 'world'," explains Giannini. "It is a world without end. . . . This spot, this point of the universe, is and always was, as definitely connected to all the rest, to the so-called 'planets,' stars and asteroids, as our arms are connected and are a part of the body."

No polar expedition yet undertaken has ever reached the poles nor flown over them, maintains the originator of a theory more daring than anything Jules Verne or H. G. Wells have ever imagined. The explorers either were mistaken in their beliefs or came to grief, like Nobile. He says:

Never Any Poles

"Let us be done with the poles. They do not exist, never did exist, except in the imagination. We are today able to move, from either the supposed North Pole or the supposed South Pole, as far as we wish—not alone for hours, but for years, into the entire vast universe, into the Cosmos, into the heavens—"

This continuation of our earth extends from the poles for millions of miles, connecting towards the north with Venus and towards the south with Mars, these planets in turn connecting with others, according to Giannini, who also claims that there is no space and that the universe has the shape of a constantly waving banner. He pictures this continuation, or trunk, as consisting of land, ice and water, with warm and frigid zones. The aviators who have become lost in the Arctic have, without knowing it, alighted on this "extension."

Columbus was warned by the foremost geographers and astronomers of his time that, should he undertake his journey, his ships would eventually come to the edge of the world and fall over. Columbus did not believe them, and discovered a new continent. Today Giannini, who is also of Genoese blood, refuses to believe that there are any poles, in the accepted term of the word. The Copernican conception once discarded, we will be able to start by easy stages to the nearest; in fact, we could even walk towards them, he says. He hopes to prove the truth of his assertions within sixty days and is now engaged in persuading a member of the prospective Byrd expedition to the South Pole to make the necessary observations in the Polar region.

Far from smiling at the enthusiastic young man, as we would smile at one who claims to have squared the circle or discovered the principle of perpetual motion, Father Ricard listened to him with close attention. At the close of the interview the Padre of the Rains voiced his opinion that the "Giannini conception" would undoubtedly prove interesting reading and set people to thinking.

Continuity

A "continuity theory" is championed by an authoritative group of scientists, including Father Ricard. However, this theory is not the one held by the Italian visitor.

The continuity theory is a matter of common sense which no man can refute, explained Father Ricard. There are no vacant spaces between particle and particle. There are no vacant spaces in the universe. He reaped the words of Captain Thomas J. J. See, of the United States Naval Observatory here, with whom he is in complete agreement:

"The earth and other planets are connected with the sun by cables of mighty force that nothing can break. This sort of connection holds throughout the universe. The least that can be said is that the ether is discharging all these functions, but more is to be admitted—namely, the play of electro-magnetism between particle and particle, between planet and planet, between planets and the sun, between the sun and all other suns of the universe, let alone the comets and planetary nebulae."

The adherents of the atomic system deny the continuity of matter, of which Father Ricard is convinced. The scholastic physicists, to whom he belongs, hold that there is a continuity between bodies even as small as modern science has been able to determine. The atomists, on the other hand, have held 'hat every atom stands separate from every other atom, and that material forces ply between these. Father Ricard and his school hold that forces are inherent in a material—and spiritual—subject, and cannot get out of them. "In the created universe there is no absolute space," he says.

Father Ricard made it quite plain that what he had in mind is a "free" continuity, and not one consisting of land and water, as Giannini claims. The Santa Clara scientist sees no reason why airplanes—or dog sleighs, for that matter—should not go over the spots we designate as "poles."

Fantastic Plan for a Trip to Mars

*A Mystic Interplanetary Bridge Will
Enable Man to Make the "Hop,"
Says This Young Magnate*

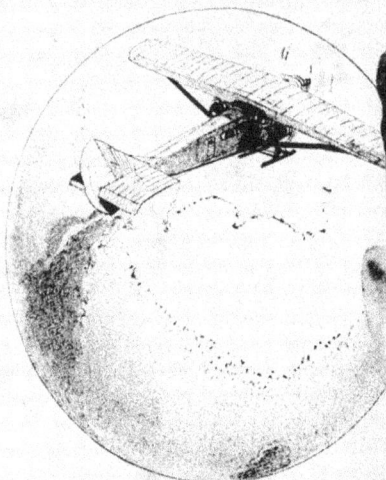

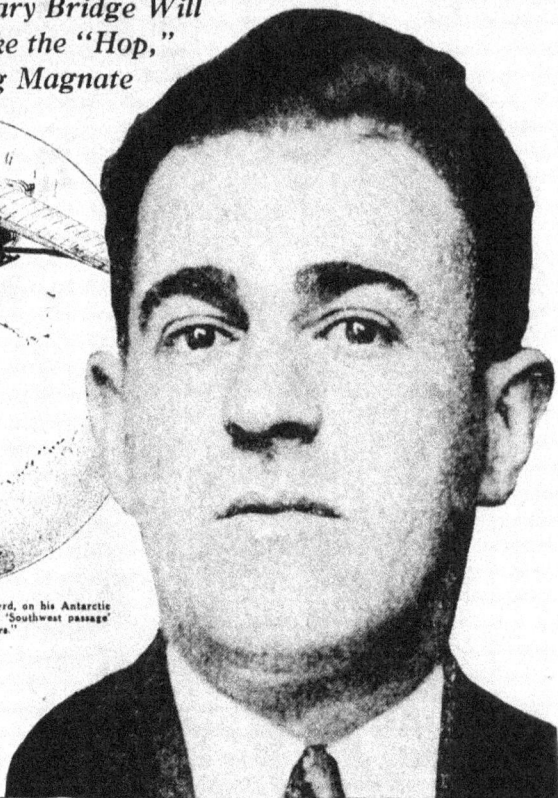

"The youthful theorist thinks that Commander Byrd, on his Antarctic
expedition, could continue on in an airplane by a 'Southwest passage'
from the South Pole towards Mars."

SPECULATIVE
Close-Up of Mr.
Giannini, Who Spends
Much Time Traveling
from Coast to Coast to Interest
Explorers and Aviators in His
"Bridge-to-Mars" Expedition Scheme.

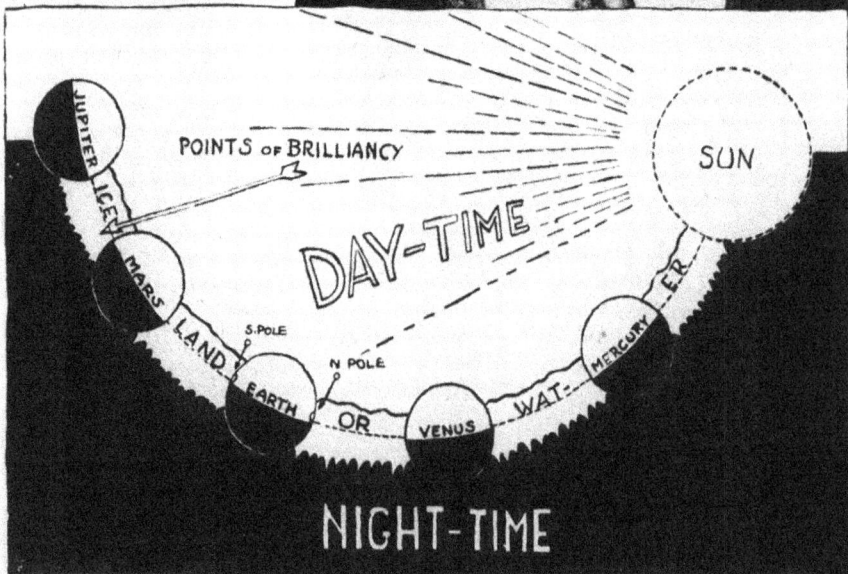

POINTS OF BRILLIANCY

SUN

DAY-TIME

JUPITER ICEY

MARS LAND

S. POLE

N. POLE

EARTH

OR

VENUS

WAT—

MERCURY ER

NIGHT-TIME

VISION OF UNIVERSE
Francis A. Giannini's Own Pictorial Conception of the Solar System, Which, He Says, Includes
the Remarkable "Interplanetary Bridge" (Not Specifically Indicated on His Diagram.)

Fantastic Plan for a Trip to Mars

A Mystic Interplanetary Bridge Will Enable Man
to Make the "Hop," Says This Young Magnate

The most novel planetary theory advanced since the days of the ancient astronomers has just been put forward — without a blush or a wince — by a young and imaginative Italo-American.

Francis A. Giannini, of New York, believes and says that man can and will reach Mars — and not by any of the old-fashioned Jules Verne expedients, either.

Instead, he forecasts, the lucky man who first negotiates the "hop" will do so by means of an existent bridge stretching from our globe to the nearest of our large neighbors in the firmament.

This will provide a stimulating thought for venturesome spirits who, having "seen everything" on earth, pant for a little faraway relaxation. More timorous[1] mortals, before setting out for a Martian holiday, however, will want to know the details of Mr. Giannini's scheme.

And Mr. Giannini is perfectly willing to tell them.

He holds that every planet in the solar system is connected with its neighbors by an endless bridge which reaches, in

[1] Timid; hesitant.

turn, to the sun. What is the composition of the bridge — ice, earth, water?

To these direct questions, Mr. Giannini has an answer pat. The bridge, he rejoined, may be of ice or earth or water, but whatever the substance, it is a solid mass extending from the poles of one planet to the next in line.

Parenthetically, he throws in the suggestion that a similar, but larger, bridge links the various solar systems throughout the universe.

Mr. Giannini's theory about the connected polarity of the planets is regarded by several people as a new contribution to scientific vision, if not to rigid scientific thought.

What, if the Giannini supposition is true, happens to persons who make polar journeys? The youthful theorist has some interesting things to say about this. He thinks, for example, that Commander Byrd[2] and the other members of his Antarctic expedition could continue on in an airplane by an as-yet unknown "Southwest passage" from the South Pole toward Mars. That is, if they wanted to and could definitely locate the passage.

Had he not met with disaster, General Nobile, too, could have continued his icy Northern voyage beyond the earth's limits.[3] Chancing on a passage similar to the Byrd one, the Italian would have eventually reached Venus.

And any day now, Mr. Giannini asserts, someone is likely to find either the northern or southern bridge, after which it will be but a matter of patience and ingenuity to quit this earth for other — and perhaps more interesting — spheres.

But there is a sinister note in Mr. Giannini's reckonings; an intimation that sounds like a quotation from an early

[2] At the time this article was published in 1928, Bryd had attained the rank of commander, but had not yet been promoted to rear admiral.

[3] Umberto Nobile (1885–1978) was an Italian aviator, aeronautical engineer, and Arctic explorer primarily remembered for designing and piloting the airship *Norge*, which may have been the first aircraft to reach the North Pole on May 11, 1926.

romance by H.G. Wells.[4] If, says he, with perfect logic, inhabitants of the earth can find a way to reach Mars, what is to prevent Martians — if any — from finding a way to reach the earth?

Nothing.

Mr. Giannini, in his more somber moments, paints a murky picture of what may happen when some Martian Lindbergh[5] chances upon the bridge connecting his home with our planet. "If there be living creatures upon Mars, similar or superior in intelligence to ourselves," speculates Mr. Giannini, "they may in time find their way to the earth. Inhabitants of this planet would then face an invasion which might result in the extermination of human civilization."

"Whoever is the first to make the discovery will obviously have the advantage in case of an invasion from the Other Shore," one of Mr. Giannini's most fervent advocates explains. "Barricades could be built; fortresses set up. Guards should be maintained, after the fashion in which Great Britain mans the Strait of Gibraltar[6] as a precaution against possible foes."

Mr. Giannini, uncertain of the precise materials making up the interplanetary bridge, looks to aviation to accomplish the feat of reaching Mars. He thinks that two generations may have passed before the attempt is successful, but that's no reason for not trying now. The actual expedition, once on its way, would consume twenty years or more before it landed on Martian soil. If further advances in the science of

---

[4] H. G. (Herbert George) Wells (1866–1946) was an English writer prolific in many genres, best remembered for his science fiction novels *The Time Machine* (1895) and *The War of the Worlds* (1898), and is often called a "father of science fiction."

[5] Charles Lindbergh (1902–1974) was an American aviator, military officer, author, inventor, and activist, who made the first solo transatlantic flight on May 20-21, 1927. Interestingly, in 1925, then Army Air Service Reserve Corps Lieutenant Lindbergh had applied to serve as a pilot on Byrd's first North Pole expedition, but his bid arrived too late.

[6] A narrow strait that connects the Atlantic Ocean to the Mediterranean Sea, separating Europe from Africa by 7.7 nautical miles (14.2 kilometers) at its narrowest point.

ch. ends p. xxxiii

flight are made the time, estimates Mr. Giannini, could be shaved in half.

Then, too, in his conception of the universe, it would be possible to effect landing, now and then, on the interplanetary bridge. "Is it more preposterous," he ejaculates, "to imagine an airplane flying by easy stages to another and not-far-distant planet than to conceive of a rocket making the journey in a single, non-stop flight?" — a crack calculated to burn up all modern rocketeers who never heard of the Giannini bridge, or if they heard of it, would think it was a novel by Thorton Wilder.[7]

But even the skeptics, unconvinced of Giannini's project, are not unimpressed by Giannini himself. For here is no flame-eyed fanatic, no foaming irresponsible, but a quiet, well-bred, faultlessly dressed gentleman, polished in speed and of an ingratiating[8] manner. That portion of his time that is not occupied with business he spends traveling back and forth across the United States in order to interview scientists, explorers, and aviators in the hope of enlisting their interest in his theory. He always stops at the best hotels.

Mr. Giannini chatted with members of the Byrd expedition shortly before it set out for the South Pole aboard the Norwegian whaling ship, *Larsen*.[9] He also talked with Sir George Wilkins[10] on the latter's latest visit to the Pacific Coast.

"I told Sir George," Mr. Giannini confided, "that if he would go to that point in the extreme South where the compass

---

[7] Thornton Wilder (1897–1975) was an American playwright and novelist who won his first of three Pulitzer Prizes for the 1927 novel *The Bridge of San Luis Rey*.

[8] Calculated to please or gain approval.

[9] The *C.A. Larsen* was named in honor (much like Antarctica's Larsen Ice Shelf) of Carl Anton Larsen (1860–1924), the Norwegian-born whaler and Antarctic explorer who made many important contributions to the exploration of Antarctica.

[10] Sir George Hubert Wilkins (1888–1958) was an Australian polar explorer, ornithologist (one who studies birds), pilot, soldier, geographer, and photographer.

needle begins to wobble and grow unreliable, he would ultimately reach Mars. You see, there is no such thing as a North Pole or South Pole. These points at which our earthly compasses cease to function are not ends, but starting places. A true compass, attuned to the center of the universe, instead of to this infinitesimal way station called the earth, would guide the interplanetary seekers to their goal."

French, German, and Russian scientists have been attracted, though not necessarily convinced, by details of the Giannini plan cabled to their respective countries. And in Southern France, a woman is actually rearing her small son to be an interplanetary explorer. The child is being trained to subsist upon rarefied[11] air and is daily instructed in the use of every system of long-range communication which in future years might aid him of "shout back" to Mother Earth once he had attained Mars.

This concern over reaching another planet has been an obsessing passion of the human race for many years. And there seems to be something fascinating to the human mind about Mars in particular. This is not only because it is so near us — it is, at its closest only 35,000,000 miles away — but because of the character of its surface markings and the probability that men, or creatures like men, may live there.

General interest in Mars was whetted by the telescopic researches of Sir William Herschel,[12] which, when published, resulted in subsequent investigations proving that the planet possessed weather conditions not entirely dissimilar from the earth's. Later the canal-like waterways observable through the telescope were thought to give evidence that Mars was inhabited; but at first this theory was subjected to severe challenge.

[11] Thin; having a low density or pressure.
[12] Frederick William Herschel (1738–1822) was a German-British astronomer and composer best known for the discovery of Uranus in 1781, the first planet to be discovered since antiquity.

ch. ends next p.

Until the middle of the 19th century, very little was known about Mars' surface, except that the white patches around the Polar regions were identified as snow, since the latter increased in Winter and decreased in Summer. Irregular, permanent, dark patches were eventually looked upon as seas, while the brighter portions of the planet's surface were labeled "land."

Reasons were adduced[13] for supposing that the atmosphere surrounding the planet was much more rarefied than ours, and less capable of being charged with clouds, unlike the sister planet, Venus, which is consistently concealed from earthly view.

Within the last century, astronomical research has developed further definite fascinating facts about the planet. Most breathtaking of these was the discovery by Percival Lowell of Flagstaff, Arizona, that the Martian canals were artificially constructed by an intelligent race of inhabitants for irrigation purposes.[14]

Scientists now generally admit that, despite the undoubted and intense Martian cold weather and the lightness of the gravity there, inhabitation is possible.

Schiaparelli, who studied the planet in 1877, mapped a network of dark lines (the canals) crossing the continental regions from sea to sea and crossing one another, some at right angles, others obliquely.[15] All were found to terminate in seas or canals. The Milanese scientist also found that at certain points the canals became double at certain seasons.

Knowledge that the amount of Martian water is small led

[13] Cited as evidence.

[14] Percival Lowell (1855–1916) was an American businessman, author, mathematician, and astronomer who fueled speculation that there were canals on Mars, and furthered theories that led to the discovery of the (dwarf) planet Pluto.

[15] Giovanni Virginio Schiaparelli (1835–1910) was an Italian astronomer and science historian who in 1887 observed a dense network of linear structures on the surface of Mars, which he called *canali* in Italian, meaning "channels," but the term was mistranslated into English as "canals."

Professor Lowell to the conclusion that the dwellers on the "fiery, red planet" had constructed the canals to conserve the supply. Seasonal changes in the dark regions, once called seas, in the southern hemisphere, are thought to be due to vegetation.

All these discoveries tended to arouse the keenest human interest in Mars and its supposed inhabitants, and the natural sequel to this has been the burning ambition to reach the planet. Scores of schemes have been incubated, trifled with, and abandoned, but the mere fact that most of them have sounded wildly fantastic has not deterred other adventurous men from putting forward plans quite as curious-sounding.

Last year, Robert Condit, of Condit, Ohio, thrilled Miami, Florida, by announcing that he would soar to Venus in a specially constructed rocket.[16] So far as is known, Professor Condit is still earthbound, but — that isn't going to deter Mr. Giannini from looking for that elusive bridge which, once found, will be a more exciting sight, a more intrepid adventure, a more heroic opportunity than the bridge Horatius defended.[17]

---

[16] Several newspaper articles were published in December 1927 reporting that Professor Robert Condit "an obscure chemist and engineer" and "ex-service man" of Condit, Ohio, did intend to fly to Venus using a machine of his own invention that would be propelled by a "slow explosive extracted from alkali peroxide of sodium."

[17] Publius Horatius Cocles was an officer in the army of the ancient Roman Republic who famously defended the Pons Sublicius (the earliest known bridge of ancient Rome, spanning the Tiber River) from the invading army of Lars Porsena during the war between Rome and Clusium in 508 B.C.

Worlds Beyond the Poles
Physical Continuity of the Universe

*"The discovery of new worlds, in matter as in mind,
is but the logical outcome of an infinite universe."*

Only Dreams are True

The tangible and real,
  On which our lives are based,
Was yesterday's ideal,
  A rosy picture traced

By some quaint visionary—
  Impractical, "half-cracked"—
Painting his fancies eerie;
  And now it's solid fact.

Whatever we hold stable,
  Dependable and sane
Was once a hopeful fable
  Of "castles built in Spain."

Before the fact, the fancy,
  Before the deed, the Dream,
That builds by necromancy
  The hard, material scheme.

So all your towers that shimmer,
  Your lamps that light the sky,
Were once a tiny glimmer
  Within some seer's eye.

Time makes our empires scatter;
  But we shall build anew,
For only visions matter,
  And only Dreams are true.

—Berton Braley[1]

---

[1] Berton Braley (1882–1966) was a prolific American poet, playwright, and author. We Heathens were unable to locate the original source of this poem as Braley was a prolific poet whose work was widely syndicated in magazines and newspapers. Our best guess is that it was first published in the 1930s, possibly with a different title.

The Changing Scene 1927-1957

1927

August

"If it is so the world will know of it."

—William Cardinal O'Connell, Archbishop of Boston

1928

July

"Giannini, since words cannot confirm you, words cannot deny you. It is your work, and only you can give it."

—Dr. Robert Andrews Millikan, President, California Institute of Technology (Pasadena)

"Giannini, if you prove your concept it will establish, the most complete Physical Continuity in the history of man."

—The Rev. Professor Jerome S. Riccard, S.J., Physicist & Seismologist, Santa Clara University (California)

December

"The memorable December 12 discovery of heretofore unknown land beyond the South Pole, by Capt. Sir George Hubert Wilkins, demands that science change the concept it has held for the past four hundred years concerning the southern contour of the Earth."

—Dumbravă, Romanian Explorer

1929

". . . Physical Continuity of the Universe more daring than anything Jules Verne ever conceived."

—*Boston American* (Hearst)

1947

February

"I'd like to see that land *beyond* the Pole. That area beyond the Pole is the center of the great unknown!"

—Rear Admiral Richard E. Byrd, U.S.N., *before his seven-hour flight over land beyond the North Pole*

1955

April 6
"Rear Admiral Richard E. Byrd to Establish Satellite Base at the South Pole."
—International News Service

April 25
"Soviet Scientists to Explore Moon's Surface with Caterpillar Tank."
—United Press

November 28
"This is the most important expedition in the history of the world."
—Admiral Byrd, *before departing to explore land beyond the South Pole*

1956

January 13
"On January 13 members of the United States expedition accomplished a flight of 2,700 miles from the base at McMurdo Sound, which is 400 miles west of the South Pole, and penetrated *a land extent of 2,300 miles beyond the Pole.*"
—Radio Announcement, confirmed by the press February 5

March 13
"The present expedition has opened up a vast new land."
—Admiral Byrd, *after returning from land beyond the South Pole*

1957

". . . that enchanted continent *in the sky,* land of everlasting mystery!
—Admiral Byrd

Figure 5

The U.S. Navy's V-2 rocket camera photographs dispel the illusion.

This stratosphere photograph of a small part of the earth's sky, taken from a V-2 rocket 65 miles up, shows the globular illusion and photographic distortion as expressed by Giannini.

The first photos taken from space were taken on October 24, 1946, on the modified, sub-orbital U.S.-launched V-2 rocket (flight #13) at White Sands Missile Range in White Sands, New Mexico. Photos were taken every 1.5 seconds with a DeVry 35mm black-and-white motion picture camera.

The highest altitude (65 miles, 105 km) was five times higher than any picture taken before.

Preface

The following pages contain the first and only description of the realistic Universe of land, water, oxygen, and vegetation, where human and other forms of animal life abound. This is not a work of fiction, nor is it a technical analysis of anything. It is a simple recital of fact which transcends the most elaborate fiction ever conceived. It is diametrically opposed to the assumptions and the mathematical conclusions of theorists and technicians throughout the ages. It is truth.

These pages describe the physical land routes from the Earth to every land area of the universe about us, which is all land. Such routes extend from beyond the North Pole and South Pole so-called "ends" of the Earth as decreed by the theory. It will here be adequately shown that there are no northern or southern limits to the Earth. It will thereby be shown where movement straight ahead from the Pole points, and on the same level as the Earth, permits of movement into celestial land areas appearing "up," or out from the Earth.

An original treatise basic to this book was written and has been expounded at American universities, 1927-1930. Since then, the U.S. Naval Research Bureau[1] and the U.S. Navy's

[1] Established by Congress in 1946, the Office of Naval Research (ONR) is an organization within the Department of the Navy responsible for the science and technology programs of the U.S. Navy and Marine Corps.

exploratory forces have conclusively confirmed the work's principal features. Since December 12, 1928, U.S. Navy polar expeditions have determined the existence of indeterminable land extent beyond both Pole points, out of bounds of the assumed "isolated globe" Earth as postulated by the Copernican Theory of 1543.[2] On January 13, 1956, as this book was being prepared, a U.S. Naval air unit penetrated to the extent of 2,300 miles beyond the assumed South Pole end of the Earth. That flight was always over land and water and ice. For very substantial reasons, the memorable flight received negligible press notice.

The United States and more than thirty other nations prepared unprecedented polar expeditions for 1957-58 to penetrate land now proved to extend without limit beyond both Pole points. My original disclosure of then-unknown land beyond the Poles, in 1926-28, was captioned by the press as "More daring than anything Jules Verne ever conceived." Today, thirty years later, the United States, Russia, Argentina, and other nations have bases on that realistic land extent which is beyond the Earth. It is not space, as theory dictated; it is land and water of the same order that comprise known Earth territory.

This work provides the first account of why it is unnecessary to attempt "shooting up," or out, from the terrestrial level for journey to any of the astronomically named celestial land areas. It relates why such attempt would be futile. These pages present incontrovertible evidence that the same atmospheric density of this Earth prevails throughout the entire Universe. Such a feature proves that, except for the presence of a gaseous sky envelope and underlying oxygen content equivalent to that of the Earth, we could never observe the

---

[2] Copernican heliocentrism is the astronomical model developed by Renaissance polymath Nicolaus Copernicus (1473–1543), which positions the sun at the center of the universe, motionless, with Earth and the other planets orbiting around it in circular paths. His model displaced the geocentric model of Ptolemy, which had placed Earth at the center of the universe.

luminous celestial areas designated as "star," or "planet." It is shown here that in a determination of realistic cosmic values the observed luminous areas of the Universe about us represent celestial sky areas, and that they are as continuous and connected as all areas of this Earth's continuous and connected sky. Hence it is shown that there are no "globular and isolated bodies" to be found throughout the whole Universe: They are elements of lens deception. Accordingly, the absence of celestial "bodies" precludes any possibility of bodies "circling or ellipsing in space."

This work is radically and rightfully opposed to astronomical conclusions of all ages. It depicts the illusions developing from all telescopic observations and photographs of the universe about us. It clearly explains and vividly illustrates why those lens-developed illusions have been mistakenly accepted as facts. The book is therefore unparalleled in the long history of man's attempted interpretation and recording of the universe about us. It projects man's first understanding of the factual and endless Universe which contains human life throughout its vast length and width — regardless of all abstract theory to the contrary.

F. Amadeo Giannini

1

Extrasensory Perception: A One-Minute Express to the Universe About Us

This is reality; it is truth stranger than any fiction the world has known: There is no physical end to the Earth's northern and southern extent. The Earth merges with land areas of the universe about us that exist straight ahead beyond the North Pole and the South Pole "points" of theory.

It is now established that we may at once journey into celestial land areas by customary movement on the horizontal from beyond the Pole points. It is also known that the flight course from this Earth to connecting land area of the universe about us which appears "up," or out, from the Earth, will always be over land, and water, and vegetation common to this Earth area of the Universe whole. Never need we "shoot up," as popular misconception demands, to reach celestial land existing under every luminous area we observe at night. On the contrary, we will move straight ahead, and on the same physical level, from either of theory's imaginary Pole points.

Confirmation of such a flight course is had in that of the U.S. Navy task force of February, 1947, which penetrated 1,700 miles beyond the North Pole point, and beyond the known Earth.[1] Additional and more recent confirmation was acquired by the flight of a U.S. Navy air unit on January 13, 1956, which penetrated 2,300 miles over land beyond the South Pole.

There is no space whatever between areas of the created Universe. But there must deceptively appear to be space in all observations. That apparent space results from the illusory globularity and isolation of celestial sky areas. The same illusory conditions have been proved to develop from observation of luminous outer sky areas of the terrestrial. "Outer sky" means the sky as it is observed against stratosphere darkness.

The concept that the Universe is comprised of globular and isolated "bodies" originated from *the curvature that is developed by all lenses.* And that lens-developed curvature fosters the deceptive appearance of globular and isolated "bodies" comprising the Universe. The "bodies" are illusory. The ancient conclusion of Galileo Galilei,[2] that luminous celestial areas are isolated from each other and are "circling or ellipsing in space" was founded on the inescapable errors of lens functioning. The "circling" movement apparent to Galileo is an illusion. In an endless land and sky Universe of reality, the undulating, or billowing, of luminous sky gas enveloping the entire Universe must deceptively appear as a circling or ellipsing movement. The deceptive appearance develops from the fact that such gaseous sky movement is detected by a circular lens. Hence there is necessarily reproduced the circular and therefore globular-appearing lens image.

Under the mobile sky gas, which extends throughout the

[1] The veracity of this claim is addressed in the article "Byrd Did Make North Pole Flight in Feb. 1947!" included in the appendices, pp. 341-358.
[2] Galileo di Vincenzo Bonaiuti de' Galilei (1564–1642) was an Italian astronomer, physicist and engineer and has been called the "father" of observational astronomy, modern physics, the scientific method, and modern science.

celestial realm, there is undetectable but very factual land, water, vegetation, and life like that common to this Earth. Therefore, the so-called "stars" and "planets" of astronomical designation are in reality lens-produced apparently globular and isolated areas of a continuous and unbroken luminous celestial outer sky surface. It envelops every land area of the celestial in the same manner that it envelops the terrestrial land.

One may question how such features were known when science was without record of them. If so, one has but to finish reading this chapter, which adequately describes how, when, and where.

It was October 1926, when he who sought the answers[3] to the Universe mysteries wandered through a woodland vale[4] of old New England, lavish with the scented breath of pine, and birch, and hemlock. There, and as if directed by some unknown force, he viewed a massed white formation of the celestial sky before it developed the luminosity which deepening twilight shadows would bring. Then it was that extrasensory perception's force was asserted, and ere[5] darkness gripped the woodland scene, the seeker in spirit viewed the vast unknown. Time and space became unknown as the portrait of cosmic reality was unfolded to this inner sight. Unmindful of the deductions and conclusions of the centuries, that formidable inner sight penetrated through the luminous sky depth of the resplendent so-called "Heavens above." Moving beyond the limited horizons of ordinary and standardized perception, he was privileged to witness that which the proud sense of sight and all its telescopic lens assistants regardless of their flaunted power, had been unable to detect from the time the first crude telescope was fashioned.

---

[3] Reminder: the "he" here is Giannini.
[4] Valley.
[5] Before.

ch. ends p. 19

The sensational portrait developed by extrasensory perception was of the sublime creative Universe pattern which had defied man's analysis from the unknown hour when terrestrial man first beheld the challenging celestial spectacle. And it brought realization that the then almost 1,900-year-old parable, "With eyes ye see not, yet believe what ye see not,"[6] should also contain the admonition that lenses patterned after the human lens will be compelled by their function to distort things and conditions, seen and supposed to have been seen, in the universe about us.

His perception's view extended a million miles and more beyond the mathematical boundaries of a fallaciously assumed "isolated globe" Earth. It penetrated through the sublime celestial domain, where deceptive lights, like flashing eyes of artful courtesans, had for untold centuries beckoned and wooed terrestrial man into their enlightening embrace. But terrestrial man, misreading the luminous signals, was denied the long-dreamed-of pleasure of their propinquity.[7] Had he properly interpreted the signals, he would have long since acquired land areas of the universe about us.

There was no misinterpretation of signals by the seeker of 1926. He journeyed to the celestial beacons on the wings of extrasensory perception's limitless necromancy. That magic permitted breaking through the long-established barriers of deduction, hypothesis, and theory. It disdainfully pushed aside the ice barriers of the terrestrial North Pole and South Pole assumed Earth ends. And there, beyond the Poles, the most fascinating creative secrets were divulged. Throughout the ages, they had been held in sacred trust for the doubter and true seeker who ventured that way. The secrets then disclosed provided knowledge of land courses into all the land

---

[6] A paraphrasing of John 20:29.
[7] Nearness or proximity; kinship.

areas of the Universe. Hence, to discerning consciousness, it was plainly shown there are no ends to the Earth.

Affliction's curse is always accompanied by a certain measure of blessing. And, alas, each blessing contains an element of curse. Hence dreamers must bear the flagellation[8] which dreams impose. Rebels must pay a price for their rebellion. They who are driven by forces obscure and extraordinary must be denied mortal contentment. Dreams that have built civilization are magnificent obsessions. But they are nonetheless obsessions; and the obsessed cannot hope to escape the ruthless whipping of obsession. The constant driving urge of one endowed with extraordinary perception demands that the substance of such perception be displayed, defended, and protected, at whatever cost. And he whose unrestrained spirit compelled the breaking of every man-made rule applying to the celestial, was forced to present his astounding findings and to make them interpretable to the majority. But that majority, accepting and abiding by the conclusions and dictums of established theory, always contentedly dwell within the safety of deduction's ordained realm, where finders and findings in the considered abnormal and fearful extrasensory realm are never welcome.

Thus how was this pilgrim from the extrasensory world to present his gifts, which were readily perceived to have originated in that fearful realm? How, at a time of midnight's darkness, was one to make plausible the brilliant light of noon to the majority who had never experienced that light? Moreover, the majority had absorbed the centuries teachings, which precluded any possibility of that light.

That which is original and is conceived beyond the limits of acceptable majority concepts need not disqualify the originator for workaday existence among the majority. For there need not be abnormality expressed in daily

[8] Beating or flogging.

ch. ends p. 19

application to demands of the social pattern. Yet the dream, the invention, the discovery, or whatever is original is too readily designated as "madness." Hence how can the originator of such considered "madness" hope to woo adherents of the organized and acceptable thing or condition which is in error? Must not the majority always consider the new course revolutionary? And if the thing or condition advanced upsets centuries of teachings, must it not be viewed as an expression of one who is "mad"?

The restless creative artist, the absorbed absent-minded inventor, the discoverer, and even the pioneer in an industrial operation may conform to the majority's social framework. But it is always a problem to introduce unwelcome findings to the majority who are absorbed in pleasing but fanciful and fallacious traditions which deny the reality of the findings.

The enduring pages of history are finely etched with record of dreamer enterprise which was diametrically opposed to the established concept of a particular time and place. But the dream helped build our civilization, despite majority disdain. It was thus from the time the "fool" threw black dirt into an open wood fire and, through such "foolishness," established the value and purpose of coal. He, and an exclusive battalion of others, represented what the majority was pleased to label "crackpots," "visionaries," "dreamers," and "madmen" all.

But they were the fearless experimenters and pure scientists comprising the always ostracized civilization-building clan. Their indomitable spirits were nourished by a creative nectar too potent for normal majority consumption. Such dreamers, forced to dwell in spacious loneliness, were with but rare exception compelled to fight alone; for it is most exceptional for members of the majority to risk their society's censure[9] by open and active cooperation with an impetuous pilgrim from the realm where dreams, so full of reality, are incubated.

The following, therefore, may serve as a timely guide for understanding values contributing toward civilization's development. And it may thereby permit easier comprehension of values this work is intended to present in terms that all may grasp. Socrates, the ancient and profound philosopher, was considered "mad" by the majority of his time and place.[10] And the immortal Christus[11] was denounced as "mad" on more than one occasion. We may read of the "strangeness" of Robert Fulton, who harbored an "insane idea" of harnessing steam for the propulsion of boats.[12] History also records Benjamin Franklin's "insane" tampering with the elements by catching lightning with his "stupid" kite and a key.

The eccentricity of Thomas Edison is, recalled. His particular "insane notion" was that of holding powerful electricity in a fragile glass bulb to produce electric lighting. Westinghouse had an equally "insane" idea of stopping a monstrous locomotive and train with nothing more formidable than the release of air: that "insanity" gave us airbrakes.[13]

Outstanding in the Dreamer's Hall of Fame is the name of Louis Pasteur. He was not a member of the medical fraternity of his time, but he contributed to medical science its most

[9] Severe disapproval.

[10] Socrates (c. 470-399 B.C.) was a Greek philosopher from Athens who is credited as the founder of Western philosophy and among the first moral philosophers of the ethical tradition of thought.

[11] "Christus" is the Latin form of the Greek word "Christos," which translates to "anointed one" in Hebrew, referring to the Messiah. In Christianity, it's primarily used as a title for Jesus Christ.

[12] Robert Fulton (1765-1815) was an American engineer and inventor who is widely credited with developing the world's first commercially successful steamboat, the *North River Steamboat* (also known as the *Clermont*).

[13] George Westinghouse Jr. (1846–1914) was an American entrepreneur and engineer based in Pennsylvania who created the railway air brake and was a pioneer of the electrical industry.

ch. ends p. 19

profound values, while followers of medical dogma were busy castigating him for such "ridiculous" enterprise and "mad" claims.[14]

This limited review of the world's so-called "eccentrics," "crackpots," and "impractical visionaries" may be continued with mention of Alexander Graham Bell's "eccentricity," his plodding perseverance provided our telephone. Telegraphy, too, was provided by the "madness" of Samuel Morse, who was guilty of the wild claim that messages can be sent throughout the world without the sound of a voice.

The entry is hardly dry on history's page recording "the Wrights' Folly"; such a term described the majority's opinion of Orville and Wilbur Wright. Yet while the normal majority ridiculed the new enterprise beyond their understanding, the Wright brothers threw tradition's restrictions to the winds and navigated the first crude airplane over Kitty Hawk.

These and an exclusive list of others who were not popular dreamed their individual dream and made that dream come true. And their particular form of compulsion was, to them, both blessing and curse.

Therefore, as we are mindful of the unchanging manner whereby Life Force at work sows perception's seeds so that mankind may always garner a crop fruitfully original, some guidance should be afforded for future reception of the seeds and the crop. Knowledge should develop that the new and the original of any time must, because of its newness and only for that reason, be decried by constituents of the old.

The old, the traditional and established, is always the sacred cow feeding on the clover of assumption in each time's pasture of cultivated and acceptable conceptional values. Therefore, it must be preserved at any cost. The new

[14] Regarded as both the father of bacteriology and microbiology, Louis Pasteur (1822–1895) was a French chemist and microbiologist renowned for his discoveries of the principles of vaccination, microbial fermentation, and pasteurization. His works are credited to saving millions of lives through the developments of vaccines for rabies and anthrax.

and unknown is always fearful to the majority. The fears attending normal pursuits within an established social pattern may be dispelled, or at least modified, by one means or another; but the fear of that which is new and unknown, and which is beyond the conditions and afflictions of the ordered pattern, must disturb the conforming majority. Routine is the order of the pattern; and though it is at times fatiguing, it embraces a measure of security symbolic of safety. Hence the new and the unknown must be in some measure resented, and must always fight for a hearing.

Human nature demands that beliefs acquired must be cherished and protected, be they ever so incomplete and faulty. "My truth is the truth, so say we all." Thus, like the porcupine projecting its quills in sensing possible danger, the majority become automatized to throw against the new and unknown the oral quills of skepticism, cynicism, and ridicule, without even hearing values inherent in the new. They fear that the new might encroach upon or upset cherished beliefs.

Accordingly, with some appreciation of guiding principles making for human concepts, we may now review the early movements of this particular work's originator in his pilgrimage to make known the unknown Universe of reality.

In the summer of 1927 this dreamer's quest led to a widely known arbiter of the mathematical Universe, a gentleman benefited with quarters in one of the famous ivy draped buildings of a New England university. After hearing only an introduction to the then unknown conception that in a realistic view of the Universe there is no "planetary isolation" and there are no ends to the Earth, the keeper of the mathematical Universe vociferously[15] exclaimed, "What! Would you have me doubt my senses?"

Tranquilly came the response: "Yes, since it is established that your sense of sight deceives you. That sense in particular

[15] In a loud and forceful manner.

ch. ends p. 19

should always be subjected to brain sight, where all true seeing is had."

The great lens manipulator knew only the mathematical Universe, and he presented it as the factual Universe. In blindness of rage engendered by fear of the unknown, he shouted, "Away with you! How dare you tell me there are no celestial spheres, and no space between such conditions?"

Undisturbed by such reception, the youthful pilgrim departed that university's magnificent halls of yearning and sought other fields for exposition of his perception's extraordinary findings. Shortly thereafter, he was graciously received in the cardinal's palatial[16] mansion at nearby Brighton, Massachusetts. There, in private audience with His Eminence William O'Connell, Archbishop of Boston,[17] an impressive word portrait was submitted of the work then known as Physical Continuum. The work was at that time most premature, for there had not been confirmation of its sensational features. Thus, when subsequently afforded press reference, it was described as "more daring than anything Jules Verne ever conceived."[18]

In that initial 1927 recital, it was shown that the theory of isolated "stars" and "planets" is founded on illusion, and it was asserted that every celestial area is definitely attached as the human legs and arms are connected with the torso. It was explained that such physical attachment of celestial areas, and the physical connections of celestial areas with the terrestrial, are always of land, water, or ice. It was further disclosed how at that time conquest of the celestial could be accomplished by penetration of land existing beyond the

[16] Resembling a palace.

[17] William Henry O'Connell (1859–1944) was an American cardinal of the Catholic Church. He served as Archbishop of Boston from 1907 until his death in 1944, and was made a cardinal in 1911.

[18] The actual quote, as per our research, was "more daring than anything Jules Verne or H.G. Wells have ever imagined." As found in: New Universe Continuity Theory is Expounded. (1928, July 29). *The San Francisco Examiner* 129(29), p. 8K.

imaginary North Pole and South Pole, or the true geographic centers of the supposedly "isolated globe" Earth. Such movement from polar areas was described as leading directly into celestial areas appearing "up," or out, from the Earth. That first day's audience with the cardinal occurred under the burning intensity of an August sun which too ardently embraced the cardinal's Brighton garden. And the Sun's warmth, in conjunction with a dreamer's dynamic recital, soon tired the aged prelate. The audience was adjourned in mid-afternoon.

On the following day, the unprecedented recital was continued with a description of what every area of the Earth's outer sky surface would present to observation from stratosphere darkness and from other land areas of the Universe. It was explained that the unified terrestrial outer sky surface would be detected as luminous and deceptively globular and isolated areas. Hence the terrestrial sky would present the identical "star and planet" pattern projected by luminous celestial sky areas.

It was then disclosed that the observable luminosity of all celestial areas results from the fact that every celestial area *possesses the same sky known to envelop the terrestrial.* It was claimed that the Earth's blue sky is luminous when observed against the dark stratosphere by inhabitants of celestial land territory. Hence it is the existence of a blue sky enveloping all celestial areas which permits terrestrial inhabitants to observe that celestial blue sky's gaseous luminosity against stratosphere darkness.

In 1927 science was without knowledge that any terrestrial sky area would be luminous when observed from beyond the sky. There had been no stratosphere observation or photography which could have shown the appearance of any terrestrial outer sky area. The first observation and photograph was achieved by the stratosphere explorer, Professor

ch. ends p. 19

Auguste Piccard, in May, 1931.[19] It only approximated a view and photograph of a terrestrial sky area from stratosphere darkness because Piccard had not achieved sufficient altitude for a completely dark stratosphere background which would properly express outer-sky luminosity.

The pilgrim who had explained such a condition as skylight had never journeyed to and within the stratosphere; yet he accurately described all that was to be seen by Piccard four years later. And his description contained all that was to be shown by the more detailed photographs procured through a U.S. Air Force stratosphere ascension over the Black Hills of South Dakota in 1935.[20] In addition to records of stratosphere cameras in 1931 and 1935, he described in minute detail that which was photographed by the U.S. Naval research Bureau's V-2 rocket cameras in October, 1946. Such photographs, procured at an altitude of sixty-five miles, showed at an oblique angle a deceptively disk-like and isolated sky area over White Sands, New Mexico, and subsequent Naval Research stratosphere photographs at greater altitudes hold most sensational confirmation of Physical Continuity.*

The unabating heat of the second day's audience at Brighton necessitated early retreat to the cool sanctuary of the cardinal's mansion, where the recital of endless worlds, and the manner of their conquest, was continued. During those hours the cardinal's black Scottie[21] was in faithful attendance. He seemed soulfully to absorb the recital's highlights;

---

[19] Auguste Antoine Piccard (1884–1962) was a Swiss physicist, inventor, and explorer known for his record-breaking hydrogen balloon flights used to study the Earth's upper atmosphere.

[20] *Explorer II* was a U.S. high-altitude two-man balloon launched from the Stratobowl in South Dakota on November 11, 1935, and reached a then-record altitude of approximately 13.7 miles.

[21] Scottish terrier.

* In another chapter is adequate explanation as to why the rocket camera of 1946 photographed a round area, as it were, "on edge" rather than the complete globe which every area of the terrestrial outer sky presents.

perhaps he wondered what a strange tale it was for such environment.

The recital described optical illusions resulting from the function of the human eye lens, and it was shown that such inescapable error of the lens had to be reproduced and enlarged upon by all photographic and telescopic lenses, which are patterned after the optic lens. It was explained how lens function demands lens convergence, and how such lens convergence produces the deceptive curvature which, in turn, is *developed by the lens* into disk-like proportion reflecting the roundness of all lenses. It was further related how lens property and function demand that every telescopically observed area of the celestial *deceptively appear to be globular and isolated.*

It was then rightfully asserted that every area of the Earth's continuous and unbroken outer sky surface would express the identical deceptions when observed and photographed from the proper altitude in stratosphere darkness and from celestial land areas. In other words, all observation of terrestrial outer sky areas from stratosphere depth and from any celestial land area would hold the illusion that the terrestrial territory is comprised of innumerable luminous and "rounded bodies," and the illusion of globularity would impose the illusion of isolation. Therefore, if the portrait produced by luminous outer sky areas of the terrestrial would be a replica of that produced by luminous celestial areas, convincing evidence would be had that astronomical observations of the celestial deals with luminous sky gases covering the celestial as they cover the terrestrial. It logically follows that the apparent globularity and isolation of celestial areas is illusion.

To use a recent but most inadequate caption by *The New York Times* (November 5, 1952), "The planets are connected."

ch. ends p. 19

The Times' account attributed such a conclusion to the California Institute of Technology.[22]

It seems fitting to note here that the author in 1928 expounded the Physical Continuum in the presence of Dr. Robert Andrews Millikan, then President of the Institute.[23]

At Brighton in 1927 the terms "stars" and "planets" were held to have meaning only for the mathematical Universe, which is based on, or developed from, the hypothesis founded on illusion. Conclusions herein related negate the existence of astronomy's "star" and "planet" entities within the bounds of reality and reason. They have application, as isolated entities, only to the world of the illusory. Thus the conclusion in a world of reality holds that such assumed entities are *lens-produced.*

It is perhaps timely to present a note for readers unfamiliar with the Copernican Theory. That theory, postulated in 1543, assumes that the Earth, as an isolated unit in space, rotated daily on an imaginary axis while prescribing a secondary motion in its yearly journey toward and away from the Sun. The theory maintains that other assumed globular and isolated areas of the Universe, the so-called "planets," likewise revolve in mathematically precise space orbits.

The concept of Physical Continuity, on the other hand, holding that the so-called "stars" and "planets" are connected luminous celestial sky areas with underlying land, requires no orbits or paths for assumed isolated areas that are not isolated. And none could be prescribed. Therefore, since such features as planetary isolation and space orbits can

---

[22] While that quote doesn't appear verbatim, he's definitely referencing: Universes Linked by Cosmic Bridges. (1952, November 5). *The New York Times* 102(34619), A29.

[23] Robert Andrews Millikan (1868–1953) was an American experimental physicist who won the Nobel Prize for Physics in 1923 for the measurement of the elementary electric charge and for his work on the photoelectric effect. He was chairman of the Executive Council of Caltech from 1921 until his retirement in 1945, and also served on the board of trustees for Science Service from 1921 to 1953.

have application only to the illusion-based mathematical Universe, any stipulation concerning Universe limitation applies only to mathematical formula. Accordingly, the earlier and concise academic expression of this work, then referred to as Physical Continuum and The Giannini Concept, reasonably opposed abstract mathematical limitations of the Universe structure.

The physical extent of the realistic Universe continues to be indeterminable, despite the sensational results of modern naval research, which brings the universe about us so much closer to our terrestrial area. Any knowableness of the end of anything presupposes knowledge of the beginning and the absurdity of abstract mathematics would be at once detected if the mathematical fraternity were to attempt designation of Creation's beginning. Though mathematics may designate a mathematical end without knowledge of the realistic beginning, such an end can hold value only for the abstract Universe of the astro-mathematician. It has nothing to do with the structure and the extent of the limitless Universe of reality.

With today's superior view of Universe reality, as acquired through research of the past thirty years, it may be gleaned that Galilean mechanics are no longer required; their purpose was to fortify the assumptive framework of the Copernican system. The laws propounded by Galileo had no consideration for then-unknown natural law which governs the realistic Universe. They had application only to that artificial Universe embraced by the Copernican formula. In the light of modern events, the premise upon which that mathematical and mechanistic Universe was erected is proved to be illusory; hence there can be no further purpose for the mechanics intended to sustain a premise of illusion.

In August, 1927, the cardinal was afforded a mental view of the polar extremities of a supposedly isolated globe Earth. Then, as the view was extended beyond the imaginary North

ch. ends p. 19

Pole and South Pole points, he observed how the polar ice barriers diminished, and they were replaced with mountain ranges, freshwater lakes, and abundant vegetation. As the voyage continued, realization came that the terrain and the prevailing atmospheric density corresponded to conditions at the cardinal's familiar Brighton estate. In that mental journey on a physical plane with the Earth but beyond the Earth, it was then understood that to reach apparent "up" areas of the celestial, one need not "shoot up." Or out, from terrestrial level: one need only move straight ahead over land continuing beyond the North Pole and South Pole points of theory.

The mental tour was directed to land underlying the luminous celestial areas astronomically designated Mars and Jupiter, where the cardinal viewed startling similarity of the terrestrial and the celestial. From such points the prelate had opportunity to observe the appearance of the approximate terrestrial sky area covering the Brighton estate. Looking up through the inner blue sky enveloping Mars and Jupiter, the cardinal shockingly beheld against stratosphere darkness countless luminous and seemingly isolated disk-like areas. They were known to be areas of the terrestrial sky, but they presented a positive duplicate of the so-called "Heavens above" as observed from terrestrial land areas. It was then realized that "up" is at every angle of observation from the terrestrial and the celestial. Hence "up" is everywhere, and it is always relative to the particular position occupied in the Universe whole. Accordingly, the "Heavens above" are everywhere.*

Twilight threw soft shadows over the cardinal's Brighton estate as we returned from the extraordinary celestial journey and the second day's audience was terminated. That journey had shown the cardinal what Galileo could

* See **Figure 1**, p. 20

not have hoped to show cardinals of his time. Galileo had been restricted to a description of only that which the *illusion producing lens* of his construction could detect. That lens was impotent to detect cosmic reality, and its successors are also impotent to detect cosmic reality. The illustrious cardinal realized the import of what had been shown. As his guest prepared to depart, he remarked, "if it is so, the world will know of it."

As the departing guest slowly trod the garden walk, where seeds of truth had been sown, the cardinal's black Scottie scampered over the green. Some of the seeds of that day's planting at Brighton were to sprout within four years, through the original stratosphere ascension of Auguste Piccard. Others required eight and twenty years, respectively, through the U.S. Army Air Corps' stratosphere ascension of 1935 and the U.S. Naval Research Bureau's V-2 rocket flight of 1946.

Contrary to popular belief, no explorer had penetrated beyond either Pole point prior to 1928. Press captions of the years have confusingly conveyed the idea that Arctic and Antarctic flights have been "Over the Pole" and therefore over the end of the Earth. Such has never been the case. Over the Pole point is possible, for there is such a mathematical point; but over the end of the Earth is not possible, for there is no end. Certain early explorers reached the Pole points, but to return they were obliged to retrace their course to the Pole point: in other words, *they had to turn around.* They did not go "over the Pole" in the manner implied by press accounts.

It is the globe symbol which conveys the false idea, for press and public, that movement "over the Pole" from one side of the Earth to the other side is possible. That symbol does not attest to the realistic extent of the Earth or the Earth's factual relation to the Universe whole. It is simply a convenience of archaic theory: it was never anything else.

ch. ends next p.

Trips from Alaska to Spitzbergen,[24] and vice versa, represent movement only in a west-to-east and east-to-west direction. They were never journeys due north from the Arctic Circle to and over the Pole. No explorer has ever moved over the Pole point, North or South, and arrived on the other side of the Earth in the manner indicated by the globe symbol.

If movement could be made "over the Pole" and it were possible to return to the starting point on the opposite side of a supposedly "isolated globe" Earth, there could be no possibility of going *beyond the Pole*, as has been accomplished since 1928. No *beyond* could exist, unless it were the originally conjectured space. The formidable factor prohibiting airplane flight, or other movement, in a northerly direction from one side of the North Pole area and arriving on the opposite side, as the globe symbol indicates, is that endless land extending beyond the Pole point. That land, unknown to the theorists of 1543, is the land this author's treatise described as early as 1927. And it is the land beyond which Rear Admiral Richard Evelyn Byrd, U.S.N., and a naval task force penetrated in February 1947.

That identical factor of land beyond applies as a prohibiting agent to any southerly movement over the South Pole which would permit return on a northerly course to other areas of the mathematically prescribed "globe" earth. All movement north from the North Pole and south from the South Pole must of physical necessity lead beyond the Earth's northern and southern mathematical boundaries. And it leads directly away from and beyond the conjectured "globe" Earth.

It should be remembered that the so-called northern and southern "ends" of the Earth were only assumed. They were never factually determined. Further, the assumptive value was imposed more than four hundred years ago, at

[24] The largest and only permanently populated island of the Svalbard archipelago in northern Norway.

a time when restrictions on polar explorations prohibited determination of factual terrestrial extent. It should also be held in mind that the Earth cannot be circumnavigated north and south within the meaning of "circumnavigate." However, certain "around the world" flights have contributed to popular misconception that the Earth has been circumnavigated north and south.

"Over the North Pole," with return to North Temperate Zone areas without turning around, can never be accomplished, because there is no northern end to the Earth. The same conditions hold true for the South Pole. All progressive movement beyond the respective Pole points leads beyond the assumed "ends" of an "isolated globe" Earth. And that area beyond constitutes a land connection with the celestial. That connecting land, though appearing "up" or out from terrestrial points other than the Poles, is attainable by movement straight ahead from the imaginary Pole points.

This is not 1927. The existence of worlds beyond the Poles has been confirmed by U.S. Naval exploration during the thirty years since then. The confirmation is most substantial, though information has not been divulged from every rostrum.[25] They of the rostrums are as little informed of the meaning of polar exploration as members of the press. That is why this book is dutifully but most arduously written.

---

[25] Podium; platform.

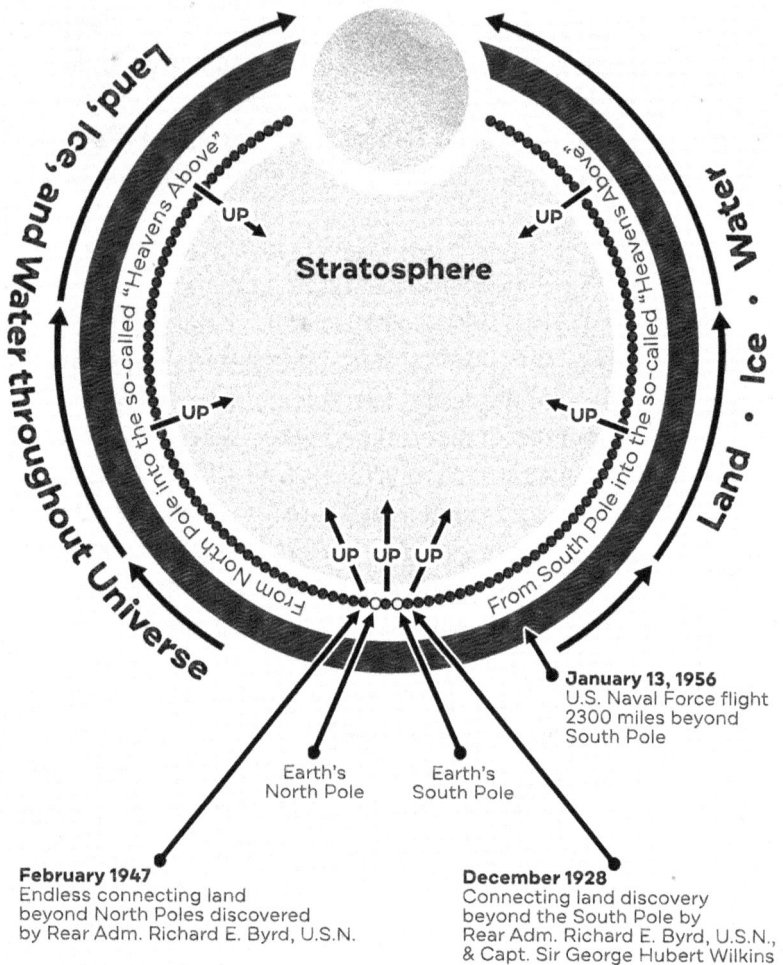

Figure 1

The universe as it must deceptively appear and as
it has been misinterpreted throughout the ages.

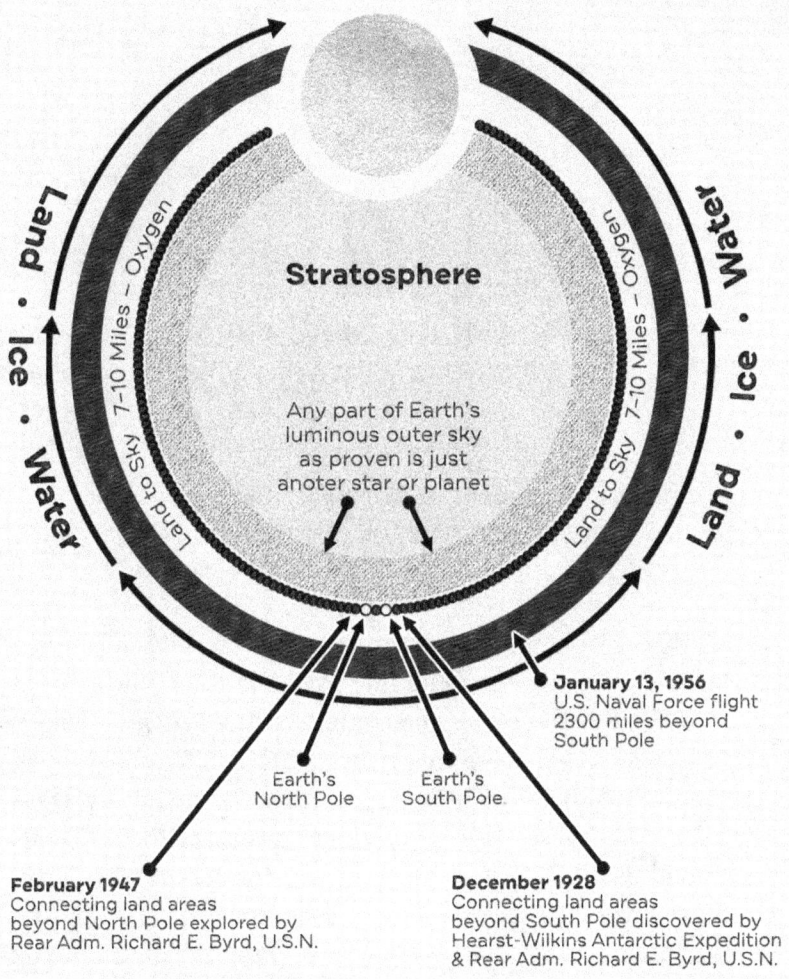

Stratosphere

Any part of Earth's luminous outer sky as proven is just anoter star or planet

Land to Sky 7-10 Miles – Oxygen

Land to Sky 7-10 Miles – Oxygen

Land · Ice · Water

Water · Ice · Land

January 13, 1956
U.S. Naval Force flight
2300 miles beyond
South Pole

Earth's
North Pole

Earth's
South Pole

February 1947
Connecting land areas
beyond North Pole explored by
Rear Adm. Richard E. Byrd, U.S.N.

December 1928
Connecting land areas
beyond South Pole discovered by
Hearst-Wilkins Antarctic Expedition
& Rear Adm. Richard E. Byrd, U.S.N.

Figure 2

The created universe as it exists on a physically connected
plane with the Earth, where every area is endowed
with identical Earth attributes.

2
The Connected Universe, Mistress of Deception

"Let us remember it is the brain that sees, and that the human eye is only a faulty window which shows us but an infinitesimal portion of the universe about us."

Figure 1 indicates the deceptions experienced in the telescopic observation of the universe about us. But it is not intended to show the true contour of the Universe whole; it is meant to express only the salient[1] features of Physical Continuity. It shows how all connected land and sky areas of the Universe have positive continuity with the Earth. But it also shows how every sky area of the Universe must deceptively appear to telescopic detection as a globular area. And that deception of globularity imposes the delusion that the areas are isolated.

Connecting areas, or parts, of the illustration's luminous outer sky curves may be considered "star" areas between the "planets." Though the illustration shows them all more or less alike, there does exist variation in their luminous depth; but they are all areas of the luminous outer sky surface of the Universe. Variations in luminous depth result

[1] Most important.

from differences in intensity of gaseous sky content. Such variations in turn develop differences in the astronomer's spectrum and spectroscopic analysis. All luminous areas of the Universe illustration are (in common with the Universe it represents) observable parts of an infinitely continuous and unbroken outer sky surface. It covers underlying celestial land, water, and ice as it covers such elements of the terrestrial.

There is also shown the region of atmospheric density between land surface and the inner blue sky. The distance is the same at celestial level as it is at terrestrial level, and the oxygen content is sufficient to sustain vegetation and life at celestial level.

In the Copernican concept of planetary isolation, the Sun is assumed to occupy the center of the dark stratosphere, and connecting outer sky areas of the Universe are assumed to be isolated units. And they are assumed to perform a rotative movement around the Sun center of a mathematically prescribed Universe subdivision known as the Solar System. That Solar System arrangement, which embraces the Earth, represents something of a combined celestial and terrestrial pinwheel. To make for easier comprehension of Physical Continuity, the pinwheel Solar System center, or Sun, has in a way been pulled out to afford it reasonable placement as a guide or leader for the entire connected Universe. As the illustration shows, every previously assumed isolated area of the Universe whole, including the Earth, holds its original position in the Universe structure, and every area maintains its daily and yearly relation to the Sun. Accordingly, the illustration shows how the land and the sky of the celestial extend to and connect with the Earth's imaginary Pole points. It shows that we may move beyond the earth without "falling off the edge" or "falling off the 'ends.'"

The following descriptive material, in conjunction with

ch. ends p. 34

the illustration, should afford ample guidance for comprehension of the factual Universe as it was created.

1. The dark center represents the perpetually dark stratosphere surrounding the terrestrial and the celestial. It is part of the dark void of infinity wherein the Universe whole was created.

2. The luminous outer partial disks, to be observed against stratosphere darkness, represent the sky-light developed over all areas of the Universe. A continuity of the same blue sky we observe from land surface everywhere on the terrestrial is seen by inhabitants of every other Universe area when they, as do we, look up or out from their respective land-surface positions. In looking through their inner blue sky at night, they observe the luminosity of our gaseous outer sky areas in precisely the same manner we observe their outer sky luminosity against the darkness. Since their lenses cannot be expected to penetrate through areas of our luminous sky-light and detect the land under our sky, it is most likely that they have deduced as erroneously of our land as we have of their land.

3. Therefore, the inner side of all outer luminous disk-like areas of the illustration may be understood to represent the familiar gaseous sky envelope observable from any terrestrial location as our particular blue sky. From all other land areas of the Universe the blue sky likewise seen represents the particular sky of inhabitants of such areas.

Inasmuch as recent U.S. Naval stratosphere photographs of outer sky areas prove them to be luminous and presenting the identical appearance of celestial areas, confirmation is had that there exists the same gaseous sky content for the celestial as is known to envelop the terrestrial. Since the luminosity of outer terrestrial sky areas corresponds to that of outer celestial sky areas, it follows that atmospheric conditions underlying the sky envelope where our celestial cousins dwell must correspond to atmospheric conditions prevailing at terrestrial level. Thus the inner blue sky must also correspond throughout the entire Universe. Our experiments show that without the existence of an inner blue sky of gaseous content there could be no luminous outer sky, which is an expression of sky gas, to be observed over terrestrial or celestial areas.

4. Hence any Martians, Venusians, Jupiterians, or Librans,[2] looking up or out from their respective land positions, are during the day permitted to view their gaseous blue sky envelope with the same varying depth, or shades of blue that we observe in our blue sky. The depth of blue will depend upon atmospheric conditions prevailing at the various celestial locations at the time of observation. Further, as the celestial sky's chemical content, or gaseous intensity, varies from time to time and from place to place, as does the content of our sky, it produces a corresponding variation in the intensity of outer sky luminosity to be observed against the dark stratosphere by remote observers everywhere. Therefore, the inner areas of the illustration, denoting terrestrial and celestial sky as observed from land surface, should not be a constant blue depth. By the same token, the outer sky luminosity will not be constant but there are variations in luminous quality. As will be later shown, variations in luminous sky movement produce, or accompany, change of blue and luminous sky expressions.

5. At night, inhabitants of all other parts of the Universe observe seemingly globular and isolated areas of our luminous outer sky in the same manner as we are permitted to observe luminous, seemingly globular and isolated areas of their sky. They are permitted to see only the outer luminous expression of our sky, as we see only the outer luminous expression of our sky, as we see luminous areas of their outer sky. Since their most powerful telescopes cannot penetrate

[2] In this context, those born in the constellation of Libra, visible in the Northern Hemisphere during late spring and early summer.

ch. ends p. 34

through our sky-light, they cannot hope to see our land or our blue sky as we see it until we arrive on the land under their blue sky. As our most powerful telescopic lenses cannot penetrate *through* sky-light of the celestial, we have been unable to detect the land and vegetation under the luminous sky enveloping the entire celestial realm.

6. Moreover, over the luminous outer surface of our entire terrestrial sky, which we know extends unbrokenly, other dwellers of the Universe are compelled to observe millions of apparently globular and therefore seemingly isolated "bodies." They are luminous sky areas, and their number would depend on the power of observing telescope lenses and other physical factors herein described.

7. Nowhere throughout the length and width of our terrestrial land and sky or throughout the endless land and sky of the created Universe do disks, spheres, or globes actually exist, despite their *seeming* existence. They are entirely lens-created; they represent the most striking examples of lens illusions ever known to man.

8. Therefore, the illustration's inner blue sky horizontal curves and the outer luminous sky curves are intended to indicate the deceptions experienced in observation. Neither the Earth nor any part of the universe about the Earth curves in agreement with the deceptions of curvature here presented. We may grant such curves realism only insofar as they have been *created by the lenses*. No lens can escape producing a curve at the proper distance on the horizontal or the perpendicular. As previously related, the physical structure and properties of all lenses demand that the curve be created. Then the lens-created curve is accentuated by concept into the full-bodied and isolated globe or sphere as distance from the photographed or telescopically observed area or object is increased. There is in reality no such curvature to the endless sky and land continuous throughout the Universe.

The only such curvature that might possibly exist, and which we could never hope to determine, would be that of a conceptional nature, *having the Universe as a Whole* curve in infinite time and space. Granting such an unverifiable arrangement for the connected Universe whole would in no way interfere with the all-important factor that the Universe is connected and continuous and that journey may be had to all areas thereof by movement on the same physical level with this Earth. That indicated movement would be straight ahead, north from the North Pole and south from the South Pole.

9. Photographs, taken whenever and wherever — in Peru, in Asia Minor, or in our own Rocky Mountains — in no way prove the so-called "curvature of the Earth." They prove only that the utilized lenses could not avoid developing curves that have been mistakenly interpreted as applicable to the Earth's contour. *The lens itself created the curvature* in the same manner that the optic lens, by grace of its structure and function, creates curves and deceptive horizons within the experience of everyone.

For example, does the sky really curve down and meet the water or the land where horizons indicate it does? We know now that it doesn't, despite appearances, because physical contact with such horizontal points proves there is no such meeting.

Does the square or U-shaped opening to a tunnel draw together, as it appears to do, and become globular to our sight as distance within the dark tunnel and away from daylight at the opening is increased? Though it deceptively appears to draw together and become globular, experience has taught us that the entrance retains its original shape and size.

Does not the square top of a brick chimney become deceptively globular as photographing altitude is increased directly over the chimney opening? Such deceptive appearance must

ch. ends p. 34

be imposed by the lens; knowledge dictates that the chimney opening does not become globular. One of the classical and most common expressions of the unavoidable deceptions ensuing from lens function is that of the two separate railroad tracks which seem to merge, or meet, in the distance.

A very modern example of illusion resulting from lens function is presented in the flight of jet planes. As the speedy jet is observed moving on a direct horizontal course from east to west, or vice versa, it must deceptively appear to be shooting up on the perpendicular, then prescribing a definite curve or arc as it approaches. Then as it passes overhead and recedes in the distance, it appears to be dropping down to the land surface. The jet's horizontal course remains the same from the time it was sighted on one horizon until it was lost to view at the opposite horizon, but the lens develops the illusion that the jet was first shooting up and then shooting down. Nothing more vividly attests that the *lens produces the curve.*

These examples, plus a thousand others that could be cited, eloquently express that all lenses are subject to the functional error of the optic lens, for all lenses were patterned after the human eye lens. This means that the lens itself, in drawing to a focal point, creates the illusory curve, and that curvature illusion in turn produces "globular" areas and objects where in fact nothing rounded or globular exists.

Therefore, as the inner blue sky seemingly dips or curves to meet the land or water, under the power of lens convergence which creates our horizons, so do the luminous outer sky areas of the Universe suffer the same affliction. It makes no difference if the sky area is of one hundred miles or of one hundred thousand miles. As distance is increased, the original illusory curve becomes deceptively filled in with body property, and there is projected the further illusion of a completely globular and isolated area. In such manner does the universe about us become cluttered with "isolated globular" and spherical "bodies" that have no part in the structure of the Universe.

In the 1931 stratosphere ascent of Professor Auguste Piccard, the photographing camera lens produced a partial disk of the terrestrial sky area which Piccard barely penetrated at an altitude of ten miles.[3] That disk development which was referred to as an "upturned disk," was partial only because sufficient distance had not been achieved from the gaseous sky area. In the subsequent 1935 Air Corps' ascent to an altitude of fourteen miles[4] there was sufficient distance from the sky area, and the partial disk became rounded-out to present the appearance of a complete disk.

One may more readily understand that lens development of curves and disks if one holds in mind a picture of the first-quarter, or crescent, Moon and mentally follows its monthly course of filling-in, or completion, to the full Moon.

Confirmation since 1935 of the unfailing development of the illusions described in all telescopic observation of the universe about us attests to the reality of Physical Continuity. Every foot of the endless celestial empire telescopically observed and astronomically designated "stars," "planets," etc., is thereby shown to be as physically connected — as **Figure 2** describes — without illusory curves. The celestial is shown to be as much a continuance of this Earth area as the various countries of the Earth are physically connected and made continuous by the known land and water links. The terrestrial has affinity with the celestial in the same manner that the States of these United States are affiliated with the national whole.

There must deceptively appear to be physical disconnections in the Universe whole, where each faultily observed celestial and terrestrial sky area, in being brought to

[3] Specifically, on May 27, 1931, Piccard reached a record altitude of 51,775 feet (15,781 meters), equal to 9.806 miles.
[4] 72,395 feet (22,066 meters), equal to 13.711 miles.

ch. ends p. 34

convergence under lens functioning, *seemingly becomes isolated* from its neighboring area — as previously described, an inescapable condition of observation. Strange as it may seem, the necessary allowances for such a handicap of observation have never been made, because the handicap, though known to be applicable to observations at terrestrial level, is denied application to observations at celestial level. Complete domination by the mathematical prescription of celestial mechanics — though that prescription contains no ingredients from the Universe of reality — has endowed illusions developed in telescopic observation of the Universe with a reality they cannot and do not possess. Therefore, we should never lose sight of the fact that the designation of celestial areas as globular and isolated is at the best a vague assumption within the world of the astro-mathematician, rather than a creative fact within the world of things of which we are a part.

With further observation of **Figure 1** one may realize that, were one occupying any area of the illustrated Universe whole or observing any area thereof from a stratosphere position, the depicted curved and luminous outer sky areas of the terrestrial and the celestial would deceptively appear as *full-bodied isolated globular entities*. This observational condition would result from the fact that when the luminous curved surface area is detected, one's mind is automatized to fill in the body proportion. In the drawing it is not possible to show the full globularity which such curved areas impose on the mind and make for the concept of isolation. Average intelligence can readily discern that the luminous curved areas will not be connected through observation. They are always disconnected. Though connected here for illustrative purposes, observation would hold a dark area at every point of connection. Thus would there develop the concept of their isolation.

Study of the inner sky curves may serve as a guide for

understanding that the lens does not conveniently prepare appearances as illustrated. The lens does just the contrary. It severs each connection; then stratosphere darkness envelops each curved area on both sides and underneath. In so doing, the area becomes isolated to all appearances. Though the inner sky curves have also been drawn as connected, the lens observing any area of the blue sky and the outer luminous sky continue unbrokenly *ad infinitum*,[5] as shown in **Figure 2**, but the lens must deny such realistic continuity.

Life is no more than our individual concept of life: we all see and believe only that which we want to see and believe. Hence "primed" observations are always of doubtful value — "as dubious as spies."[6] Nonexistent celestial globular and spherical areas are clothed with reality through the capriciousness of optic lenses, aggravated by other lenses, and conceptional enlargement of the faulty image. So long as one observes luminous celestial and terrestrial sky curves produced by the lens and holds the illusory globe to be reality, it is unlikely that anything but globes and spheres will be encountered, regardless of the power of telescopic lenses. Moreover, the assumed Earth sphere and its companion celestial pseudo-spheres have become so firmly fixed in mind that presentations of such spheres, which naturally show full-bodied properties, are accepted as being factually descriptive of the composition of the Universe.

Such acceptance is had in spite of the overpowering fact that *no telescopic observer and no photographing camera ever recorded realistic body proportions for any area of the Universe.* The lenses detected and reproduced only a disk-like surface area which was credited with body fullness. Therefore, the glamorously portrayed Earth globe and its

[5] Translated from Latin: to infinity; forevermore.

[6] Likely quoted from the 1949 true crime book *Verdict in Dispute* by English author and broadcaster Edgar Marcus Lustgarten (1907–1978), wherein the following is found on p. 21: "It is notorious that people see what they expect to see; primed observers are as dubious as spies."

ch. ends p. 34

celestial counterparts present nothing more profound than an outstanding expression of lens error and human misconception based on that error, plus the artful embellishment of globe symbols by otherwise capable artists who likewise are under the domination of the popular misconception.

Modern discovery establishes that the assumed isolation of the terrestrial from the celestial is a fallacy. The Earth's northern connecting land link with the celestial is confirmed point of theory. In February, 1947, a United States Navy Arctic expeditionary force, under the command of Rear Admiral Richard Evelyn Byrd, achieved a memorable seven-hour flight over land extending beyond the northern geographic "center" or mathematically prescribed northern "end" of the Earth. That flight confirmed that there is no northern physical end to the Earth and that the 1543 conclusions were most premature. The northern Physical Continuity of the Earth with celestial areas of the Universe also has its counterpart in the land now known to extend beyond the South Pole.

All future physical progress beyond the imaginary North Pole and South Pole points must and will lead into real land areas of the Universe appearing "up," or out, from our present terrestrial position. We may move, as Rear Admiral Byrd moved, beyond the North Pole and out of physical bounds of this Earth, on the same physical level as this Earth. Our movement into land areas of the universe about us need never vary from known movement in journeys from New York City to Chicago, or from Boston to Hong Kong, or between whatever terrestrial points one is pleased to consider. We may fly the distance with means now at hand, or we may journey in any of the other established modes for making possible journeys from city to city and from nation to nation of the terrestrial area.

Except for the vast ice barriers at the Arctic and Antarctic regions, especially at the Antarctic, we might even walk. However, early explorers found walking and dog-sled movement most unsuitable transportation over frigid polar areas. That is one of the reasons why there was no concerted early effort to peer "over the top of the world," so called, to determine what actually exists beyond the supposed terrestrial ends. Further reason may have been that one does not perilously attempt to penetrate into a beyond which his concept denies. If concept has not first established the thing or condition — in this case land beyond the Poles — it cannot and does not "exist," despite its reality.

Despite the lamentable[7] restrictions of theory, men have persistently wondered about the Earth's extent.[8] The earliest attempt to reach the North Pole point and to satisfy that curiosity was made by Sir Martin Frobisher, of England, in 1578.[9] But the notable accomplishment of just reaching the Pole point could in no manner permit determination of territory extending beyond the Pole point and out of bounds of the theorized Earth "globe." One is not permitted a vista of polar territory to almost unlimited horizons as one is in viewing the plains of Kansas. One's determination must be based solely on the mathematical formula which maintains that the geographic point reached is in fact the end. And though infinity extends beyond in a continuous land and water course, men would have no reason or inclination to penetrate that course if concept holds that such course did not exist. Therefore, though the space myth did not restrict movement to the polar areas of an assumed Earth ending, it did most definitely restrict movement beyond such supposed Earth ends, where men believed they would be projected into space assumed to exist beyond the ends.

Hence the awesome conjectured northern and southern

[7] Deplorable or unsatisfactory.

[8] The range or distance over which something extends.

[9] Sir Martin Frobisher (c. 1535–1594) was an English sailor and privateer who was an early explorer of Canada's northeast coast while pursuing the Northwest Passage (the sea lane between the Atlantic and Pacific oceans through the Arctic Ocean).

ch. ends next p.

space of the Copernican Theory erected the identical barriers to northern and southern progress as the obsolete Ptolemaic Theory[10] had imposed on movement east and west from the Old World prior to 1492. How fearful has been the word "space"!

---

[10] Developed by Alexandrian mathematician, astronomer, astrologer, and geographer Claudius Ptolemy (c. 100-170; and the P is silent) from the hypotheses of earlier philosophers, the Ptolemaic System is the theory that the earth is the stationary center of the universe and all celestial bodies revolve around it.

3

A Modern Columbus
Seeks a Queen Isabella

Returning to the 1927–28 quest of the pilgrim to whom the chapter title refers (and as press accounts of that time described him), we can review his lonely pilgrimage from the cardinal's Brighton mansion. Along a lonely homeward course and in the disturbed vigil of ensuing months and years, he was taunted by the cardinal's parting words: "If it is so, the world will know of it." Silently, but no less firmly, he answered their thudding echo: "Yes, my cardinal; it *is* so. And, by God, the world *will* know of it through my telling. For I will tell, though Earth and Hell oppose me."

He could not then foresee that the combined forces of life would weave the pattern of his movements so that he must tell even though it beggars him of all worldly values and leave him outcast in the eyes of men. He was not to be consulted by the force that relentlessly drove him forward. And if he tried to escape the burden of responsibility, as try he did at times, he was mercilessly scourged by the mean expressions of "man's inhumanity to man" in fitting compensation for his periodic forlorn attempts to abandon his endowment.

There were none in whom a dreamer so endowed could confide. Alone, he was compelled to chart the forsaken pilgrimage leading to his avowed goal of universal dissemination of his work and its ultimate confirmation. Where would he go? To whom could he and would he divulge the devastating secrets culled from the hidden depths beyond accepted standards of perception?

In any really determined quest for light, a beacon, be it ever so feeble, throws its ray to guide the seeker's course. Hence, there was brought to that early pilgrim the name of one who, though serving the interests of the traditional and the entrenched, was by no means lacking in perception. To him, in the District of Columbia, the quest was conducted.

Arriving at the national capital, the pilgrim hastened to the offices of Science Service,[1] where he met with one of the few open-minded men of science. With such open-mindedness he was able to perceive beyond the established pattern of cosmological values. Dr. Edwin E. Slosson, then the fearless Director of the Science Service,[2] patiently listened to a dramatic recital without parallel which described how one might journey straight ahead from the supposed Earth "ends" to arrive at celestial land areas, how movement up is always relative, and apparent "up" points of the Universe would be attained by moving straight ahead in a manner comparable to the western sailing of Christopher Columbus to go to the East. Dr. Slosson was not an astronomer, nor was he afraid of space phantoms. However, though he fully grasped the import of sensational disclosures, he was obliged to counsel, "Giannini, you will not find ten open-minded men of science throughout this entire country."

Despite such sincere counsel, ten men of tolerance were thereafter ardently sought. It mattered little to the pilgrim whether they bore the label of "scientist" or something else.

[1] Society for Science, founded as Science Service in 1921, and headquartered in Washington, D.C., is a non-profit organization dedicated to the promotion of science and expanding scientific literacy through its science education programs, publications, and scientific research.

[2] Edwin Emery Slosson (1865–1929) was an American magazine editor, writer, journalist, chemist, and the founding director of Science Service.

If they existed and could assist in the cause, they should be found. Zeal born of relentless obsession would tolerate no cessation of the quest, which was expected to develop the means for adequate disclosure and ultimate confirmation of reception's extraordinary findings. He realized at an early date in the pilgrimage that expensive stratosphere ascent and elaborately equipped expeditions beyond the North Pole and the South Pole would be required for essential confirmation of his disclosures. And with such realization he was painfully aware that he was a dismal pauper, according to this world's standard of values. He had no way of knowing then that his utmost wish would be gratified through the physical initiative of others who would see to it that confirmation would be developed. The required stratosphere ascent and expeditions would be made.

Though he would have willingly risked his life in pioneering stratosphere ascension to procure proof and in a dangerous journey to land he knew continued beyond the North Pole and the South Pole, his earnest appeals for adequate financing of such projects fell upon deaf ears. Never relinquishing the idea of immediate physical confirmation of his disclosures and the manner of its attainment, he journeyed to California, where, at the California Institute of Technology,[3] he met that institution's president, Dr. Robert Andrews Millikan. He believed that Dr. Millikan, who had then recently accomplished isolation of an electron and was acclaimed the world's outstanding physicist, would be endowed with the open-mindedness necessary for a program developing confirmation of the extraordinary disclosures.

The famous physicist graciously afforded the hearing that presented pertinent features of the original treatise, *Physical Continuum*, also known as **The Giannini Concept**. There was

[3] The California Institute of Technology (branded as Caltech or CIT) is a private research university in Pasadena, California, and due to its history of technological innovation is considered one of the world's most prestigious universities.

ch. ends p. 46

no doubt concerning Dr. Millikan's interest. Yet his counsel and only contribution to the cause was expressed in the following: "Giannini, it is your work, and only you can give it. Since words cannot confirm you, words cannot deny you. My best wishes for your success." His words, in that remote summer of 1928, were certainly friendly and well-intended; but to the lonely and unaided pilgrim they held a dismal echo of the preceding summer's dictum from the cardinal's mansion: "If it is so, the world will know of it."

"If it is so, the world will know of it." "Giannini, you will not find ten open-minded men of science in this entire country." "Giannini, it is your work and only you can give it." In his youthful enthusiasm, he became scornful of the lack of constructive initiative from arbiters of the established order of things scientific.

Throughout the weary pilgrimage of years, a thousand and one clutching tentacles of despair sought throttling hold upon his spirit. Alone, with the soothing balm of Arizona's silent and spiritual desert nights, where he had temporary sanctuary, he often whispered a devout prayer of attunement to that Inscrutable Force which guided a dreamer's destiny: *"Padre mio! Padre mio!*[4] Show me the way!"

Then it would seem that the myriad beacons of the desert sky would direct his course back to California, to that fabulous land of the setting Sun where there seemed to remain some remnant of the pioneering spirit in keeping with broader horizons. There, where miracles of nature's vast performance tax credulity, it was believed there might be less of that finely developed cynicism infesting eastern metropolises, "whose lights had fled, whose garlands dead," and where dreams had been long *verboten*.[5] It was hoped

---

[4] Translated from Italian: My father! My father!
[5] A German word meaning forbidden.

there might be found the sordid but necessary means for dream's fulfillment through the cooperation of the master financier, Amadeo Peter Giannini,[6] who had then recently endowed the Giannini Agricultural Foundation at the University of California with two and a half million dollars.[7]

Whatever his hopes may have been, it was enough that the land of the Golden Gate had beckoned. The pilgrim proceeded to San Francisco. Then in a rapid series of events during the remainder of 1928, his work was expounded before faculty members of the University of California at Berkeley,[8] at Santa Clara University in Santa Clara's bountiful valley of orchards,[9] at the San Jose State Teachers' College,[10] at the United States Naval Observatory on Mare Island,[11] and at the Archbishop of San Francisco's headquarters, where His Excellency Archbishop Edward Hanna presided.[12] Little time was lost in an itinerary that subsequently took him to Los Angeles, where his treatise *Physical Continuum* harshly invaded the University of Southern California[13] and the

[6] Amadeo Pietro Giannini (1870–1949) was an Italian-American banker who founded the Bank of Italy (Bancitaly), which became Bank of America.

[7] The Giannini Foundation of Agricultural Economics was founded in 1928 from a $1.5 million gift to the University of California from the Bancitaly Corporation in honor of Giannini, and is used to promote and support research on the economics of California agriculture.

[8] Established in 1868, the University of California, Berkeley, is the state's first land-grant university and the founding campus of the University of California system.

[9] Established in 1851, Santa Clara University is a private Jesuit university in Santa Clara, California, and the oldest operating institution of higher learning in the state.

[10] Established in 1857, San José State University (known as San Jose State Teachers College from 1921 to 1935) is a public university in San Jose, California, and the oldest public university on the West Coast.

[11] Mare Island is a peninsula in Vallejo, California, about 23 miles (37 kilometers) northeast of San Francisco.

[12] Edward Joseph Hanna (1860–1944) was an American prelate of the Roman Catholic Church, who served as archbishop of San Francisco from 1915 to 1935.

[13] The University of Southern California (USC) is a private research university in Los Angeles, California. Founded in 1880, it is the oldest private research university in California.

ch. ends p. 46

University of California at Los Angeles.[14] It was later heard by prominent representatives of the Hearst organization,[15] who were then preparing for the historical Hearst-Wilkins Antarctic Expedition of 1928.[16] His unquenchable ardor was manifested in every quarter where his cause might be advanced. He was heard in restricted academic circles as well as in weekly lectures from Los Angeles radio station KFI.[17] He was invited to accompany Captain Sir George Hubert Wilkins and Alan Lockheed, President of the Lockheed Corporation,[18] to a select meeting at the Breakfast Club in Burbank,[19] where his cause was heard. Wherever it was considered that the work's interest might be served, he was to be found.

It is understandable that a press dispatch of that time described him as "the modern Columbus who seeks a Queen Isabella[20] somewhere in America."[21] Though a queen might have possessed the means to equip a fitting expedition for land discovery beyond the Poles or to provide funds for the required stratosphere ascents, no queen, duchess, or bar-

[14] The University of California, Los Angeles (UCLA) is a public land-grant research university in Los Angeles, California. Established in 1881, it is the second-oldest of the ten-campus University of California system.

[15] Hearst Communications, Inc., often referred to as Hearst, is an American multinational mass media and business information conglomerate founded in 1887 by William Randolph Hearst (1863–1951).

[16] After Sir George Hubert Wilkins' April 15-16, 1928, trans-Arctic crossing from Point Barrow, Alaska, to Spitzbergen, Norway, William Randolph Hearst became financier of Wilkins' polar explorations.

[17] KFI AM 640 debuted on April 16, 1922, and was the first U.S. station west of Chicago to broadcast at 50,000 watts.

[18] Allan Haines Lockheed (1889–1969) was an American aviation engineer and businessman, who formed, along with his brother, Malcolm, the company that became the Lockheed Corporation.

[19] The Los Angeles Breakfast Club is a nonpolitical, nonsectarian social club founded in 1925.

[20] Isabella I (1451–1504) was Queen of Castile and León from 1474 until her death in 1504, and is most remembered for enabling Christopher Columbus's voyage to the New World.

[21] The actual quote is "...young Italian scientist, who seeks a Queen Isabella somewhere in America" as read in: Poles Not at Earth's End, S.F. Told. (1928, June 29). *The San Francisco Call and Post* 123(147). A1.

oness ventured forth to ease a modern dreamer's burden. It appeared that modern queens and lesser members of nobility were too sophisticated to be intrigued by a dreamer's announcement of new worlds to conquer.

However, the dreamer and the dream did not perish for want of queens, duchesses, or other noblewomen. It was evident that a more alert nobility was to be found in San Francisco, for it was there that a ranking member of the Church nobility, in the person of Archbishop Edward Hanna, made possible a hearing of the pilgrim's work by the faculty of the University of Santa Clara. The famed Jesuit, the Rev. Jerome S. Ricard, S.J., who was popularly known as "the padre of the rains" as a result of his accurate weather predictions, was perhaps the most-interested member of the faculty audience.[22] His interest would rightfully surpass that of the pure academician, because he was an atomic physicist and seismologist. When the hearing was over, Professor Ricard exclaimed with undisguised enthusiasm, "Giannini, if you succeed in proving your concept of Physical Continuum, it will represent the most realistic physical continuity of the Universe within the history of man."

Professor Ricard's teachings held that there existed a constant play of energy between all assumed "bodies" and particles of the created Universe whole. However, his dignified membership in the order of theorists adhering to the supposition of 1543 did not deny him discernment that the four-hundred-year-old theory failed to provide an answer to the Universe riddle.

The *San Francisco Call*[23] of that time featured an exclusive interview with the pilgrim whose extraordinary disclosures had been made at Santa Clara University. The press

[22] Jerome Sixtus Ricard, S.J., (1850-1930), known as "Padre of the Rains," was a professor at the University of Santa Clara, who published a monthly magazine, *The Sunspot*, which became famous due to his Pacific Coast weather reports that were 99.07% accurate.

[23] A San Francisco newspaper founded in 1856, which ceased publication in 1965 after being purchased by the *San Francisco Examiner*.

ch. ends p. 46

presentation contained the pilgrim's photograph with that of the Australian explorer, Captain Sir George Hubert Wilkins. There was also a likeness of the ancient astronomer Copernicus, reproduced from an old woodcut. The feature dealt with Sir Hubert's then forthcoming Antarctic expedition, to discover unknown land beyond the South pole point.[24]

Yet even that timely and most sensational presentation failed to bring forth a queen or a duchess, or even a lowly baroness, to lend oil for a dreamer's turbulent and engulfing waters of workaday application to his dream's dissemination. As there was a notable dearth of queens and their noble retinue,[25] kings of finance and members of their noble American order were also *in absentia*.[26] No subsidy was to be had from the famous banking house of Giannini, though its master, Amadeo Peter Giannini, had been given personal knowledge of the dream's import. However, it must in fairness be acknowledged that his friendly reception, and his expressed willingness to cooperate in other than a financial way, held a measure of aid which was perhaps greater than any financial disbursement for the cause. Nor was there any assistance from the vast storehouse of private funds for the express purpose of advancing science in all its branches, regardless of scope. The overlords of that storehouse expressed the utmost skepticism concerning the land which a dreamer knew existed. One of the few cooperative courtesies of the time was extended by the United States Navy, through its senior professor of mathematics who was also Director of the U.S. Naval Observatory on Mare Island, California. He graciously permitted observations to be made with naval equipment. Though more substantial and direct aid was then withheld by the Naval Research Bureau, there was an extravagance of indirect aid which was never anticipated. This

[24] See facsimile at right: Mars Just Easy Walk Away. (1928, August 30). *The San Francisco Call and Post* 124(43). A3, A7.
[25] Advisers, attendants, or assistants to an important person, royalty, etc.
[26] Translated from Latin: in absence.

'WALKING TO MARS' IT CAN BE DONE! SO OPINES LOCAL MAN

LEAVING HIM FLAT
Copernicus, the ancient who blasted theory that earth was flat is in turn blasted today by a San Francisco young man who claims to have found universe to be "continuous."

THE RAGGED EDGE
Civilized folk, with the exception of Wilbur Glenn Voliva, have long scouted this sort of happening—that of a man falling over the edge of a "flat" world.

ALL AN ENDLESS CHAIN
Here is Francis A. Giannini, cousin of the famed "A. P." of this city, who claims he could walk to Mars on a path which begins at the South Pole, a route of ice, land and ocean.

POINTS of BRILLIANCY

JUPITER · ICE · MARS · LAND · EARTH · S. POLE · N. POLE · OR · VENUS · WAT-ER · MERCURY

DAY-TIME

SUN

NIGHT-TIME

HERE IT IS
Here is an artist's version of the endless road leading from our poles with sun's light reflected by brilliant rays.

MARS JUST EASY WALK AWAY

"An airplane could fly from Mills Field to the planet Mars. Or, if the aviator had to pawn his plane, he could walk to the said planet.

Francis A. Giannini of New York, who says he is a second cousin of A.P. Giannini, the famous banker, and who is visiting in San Francisco, is the man who makes this statement.

Giannini has told Sir George Wilkins that if he will just hold steady and not lose his nerve he can fly right on to Mars from the South Pole.

Giannini is ready to supply Sir George with all sorts of data on his 'continuous universe' theory, with a view to throwing actual light on his entirely revolutionary idea of astronomy.

IT MAY BE VALUABLE ADDITION TO HIGHWAY SYSTEM
Lower pictures show an old view of Columbus taking leave of his rulers to find new world, and Capt. Sir George Wilkins, explorer, Antarctic bound, who, Giannini says, may be a second Columbus. Local man believes Wilkins, who will fly over South Pole like Byrd, could blaze an airway to Mars if he kept going in a southerly direction after crossing the world tip.

volume attempts to describe the sensational accomplishment of record, since 1928, by the Navy's technical and explorative divisions and the Naval Research Bureau.

Though the interests mentioned here were perhaps rightfully reticent of openly assisting, in view of seemingly fantastic aspects of the Physical Continuum before confirmation, it was also rightful for their attitude to be resented by one who as yet had no awareness of the magnitude of his disclosures. To him, they were of utmost simplicity. Therefore, it may be that in the sublime unfathomable order of things this particular dreamer was, even against his wish, safeguarded from the dangers attending his desired stratosphere ascent and hoped-for flights beyond the Poles. Had he then possessed knowledge of coming events, he might not have considered it so imperative that he personally perform what he considered necessary for confirmation of his revolutionary disclosures. He lacked such knowledge, and the factor of personal safety never entered his calculations.

He sought all possible understanding of balloon construction and operation, and he solicited the cost of balloon material for the stratosphere ascent he was positive would develop proof for his unorthodox claims. He determined the cost of stratospheric balloon equipment from the Thompson Balloon Company of Aurora, Illinois.[27] He received the promise of Captain Ashley C. McKinley, U.S.N. (Retired), to pilot the ascent. Captain McKinley was then an aerial photographer who had been an expert naval balloonist.[28]

Then his earnest petition for necessary funds to procure equipment was denied by no fewer than four prominent

[27] Operating from 1903 to 1939, the Thompson Brothers Balloon & Parachute Company manufactured and distributed balloons, parachutes, and aerial apparatus for performers, and also offered parachute leaps from airplanes and balloons.

[28] Ashley Chadbourne McKinley (1896–1970) was a colonel in the U.S. Army Air Corps and an accomplished American aerial photographer, who helped pioneer aviation at subzero temperatures. As an aerial photographer, he accompanied Richard E. Byrd during his South Pole expedition.

millionaires to whom he had personally appealed and who had previously expressed intention to cooperate. Thus until 1935 he persisted in forlorn endeavor to have his own stratosphere ascent financed. At the Transamerica Corporation,[29] in New York City, he again met with the famous A. P. Giannini, whose problems of that time left him unreceptive to the stratosphere project.

His devotion to the cause actuated a journey to the Chicago World's fair, where he consulted with Dr. Frank Moulton,[30] Director of the Science Division, for stratosphere ascension to be launched from Soldier Field.[31] However, it developed that Commander Settle, U.S.N.,[32] had already been assured of *Chicago Daily News*[33] support for his stratosphere ascension. Therefore, the pilgrim, denied his own ascent and fully convinced that Commander Settle would not achieve sufficient altitude for photographic proof, took advantage of every opportunity to influence others who were favored by organization financing and who might be able to procure requisite confirmation. It was with such in prospect that he arranged an invitation to inspect the Army Air Corps' stratosphere ascension equipment at Wright Field, Dayton, Ohio.[34] And it was there that he directed Captain Albert W. Stevens, U.S.A., to achieve a fourteen-mile altitude

[29] Transamerica Corporation is an American holding company for various life insurance companies and investment firms operating primarily in the United States.

[30] Forest Ray Moulton (1872–1952) was an American astronomer and professor at the University of Chicago, whose primary interests were in the application of mathematics to problems in astronomy. In addition to writing several popular texts, he had a regular radio science program, and was one of the first professors to try to use the medium of radio to popularize science.

[31] Opened in 1924 and reconstructed in 2003, Soldier Field is a multipurpose stadium on the Near South Side of Chicago.

[32] Thomas Greenhow Williams Settle (1895–1980) was an officer of the United States Navy, who, together with Army major Chester L. Fordney, set a world altitude record in the *Century of Progress* stratospheric balloon on November 20, 1933.

[33] An afternoon daily newspaper published between 1875 and 1978.

[34] Established in 1917 as a military installation for use during World War I, Wilbur Wright Field is, today, part of Wright-Patterson Air Force Base.

ch. ends next p.

if it was physically possible.[35] He then knew that such altitude would be required for photographic confirmation of terrestrial sky-light and the illusory globular and isolated appearance of any sky area photographed.

In the case of polar expeditions to confirm his disclosure of then unknown land existent and extending beyond both Pole points, it was considered imperative that some known explorer of polar areas be convinced of the reality of Physical Continuity. To that end he determined to present the subject to Captain Sir George Hubert Wilkins, who at that time (September, 1928) was about to embark upon the Antarctic expedition sponsored by the Hearst newspaper interests.

[35] Albert William Stevens (1886–1949) was an officer of the United States Army Air Corps, balloonist, and aerial photographer. In 1930, he was the first person to photograph the Earth in a way that the horizon's curvature is visible. Then, on November 11, 1935, he commanded the record-breaking ascent of the *Explorer II* (see footnote #13, p. 12). The photograph shown above was taken during that ascent, at an altitude of 72,395 feet (13.7 miles/22 kilometers), and the horizontal black line spanning the entire photograph represents the level horizon, with Earth's curvature appearing slightly above that line and tapering down at the edges.

4

Disclosing the Southern Land Corridor into "The Heavens Above"

The pilgrim of 1928 accompanied Captain Sir George Hubert Wilkins to a meeting of the Los Angeles Breakfast Club, where Sir Hubert was guest of honor. And he later visited with the famous Australian explorer at his quarters in Hollywood's Hotel Roosevelt,[1] where the salient features of Physical Continuity were illustrated with a miniature globe symbol that permitted the quadrants of the globe to be detached. Needless to relate, greatest stress was laid on the feature of terrestrial land extent. Sir Hubert was fully informed of the unknown and endless land extending beyond the South Pole point, where his expedition was directed.

That conference was of somewhat different nature from some others of this chronicle, for the "modern Columbus" was being heard by one who was also a dreamer as well as a courageous performer in the world of established reality. Hence, the archaic of theory was not permitted to dominate the conference.

It became evident that the explorer was not risking his precious life at the forbidding South Pole merely for the

[1] Built in 1926, the Hollywood Roosevelt Hotel is located at 7000 Hollywood Boulevard in Los Angeles.

purpose of measuring wind velocity and to gauge the directional activity of ice floes. Sir Hubert seemed wholeheartedly to share the conviction that the South Pole was by no means the southern end of the Earth. His statement afforded eloquent testimony that he was possessed of a powerful urge to go beyond all restrictions of theory in the pioneering spirit of a true explorer: "You know, before leaving England I was advised that if I succeeded in penetration beyond the South Pole point I would be drawn to another 'planet' by the suction of its movement." That provided appropriate amusement, in view of the perceptional portrait then being exhibited. Yet they who were responsible for such expression were not to be censured, the Copernican concept, holding the Universe to be comprised of "isolated globular bodies," permits no other conclusion than that space would be encountered beyond the Pole points of theory.

Sir Hubert was visibly impressed by the prospects presented, and he gave firm assurance that he would continue beyond the traditional mathematical end of the Earth when he said, "Giannini, if you will show me the route to the land you claim exists beyond the South Pole, I will continue on to it in spite of all obstacles." The International News Service[2] at Los Angeles received copy of information designating the route requested by Sir Hubert. And history records his memorable discovery of land beyond the South Pole on December 12, 1928.

The manner in which the theorists may have thereafter misinterpreted the value of that land has very little meaning for this work, dealing with cosmic reality and diametrically opposed to the conjectures of theorists. However, it seems fitting to here reiterate that man's habitual fear of the unknown permits gross misinterpretation of values

[2] Operating from 1909 to to 1958, the International News Service (INS) was a U.S.-based news agency founded by newspaper publisher William Randolph Hearst.

demanding a change of concept. Man hates to forsake the old and known course. Though newly discovered facts establish that the cherished old of theory has no application to a world of reality, only with the greatest reluctance is the old relinquished.

Accordingly, there was early evidence that such previously unknown land beyond the South Pole was being subjected to a mathematical disguise which was intended to hold intact and preserve the four-hundred-year-old conjecture. The theory was not modified to fit the fact of land extent; but the land extent was discounted to make it fit the theory. The reason and purpose for that southern land extension, linking our Earth with the universe about us, was obscured with another patch of mathematical abstracts generously applied by the theorists. They served only to make glaringly ridiculous an issue which was then confused out of reason's bounds.

Therefore, it is still of timely value to quote another fearless dealer in reality who was heard immediately after Sir Hubert's memorable land discovery of December 12, 1928. The masterful arbiter of fact was the then famous Romanian explorer Dumbravă,[3] who announced, "The sensational discovery of land beyond the South Pole by Captain Sir George Hubert Wilkins, on December 12, 1928, demands that science change the concept it has held for the past four hundred years concerning the southern contour of our Earth." Dumbravă, in common with Sir Hubert and a very select group of that time, was unafraid of the space phantom projected by theorists. And, as his words expressed, he had no patience for the fearful mathematical patchwork to provide a feeble temporary, but grossly contradictory, explanation of that previously unknown land's existence.

Although the extent of that southern land continuity

[3] Constantin Dumbravă (1898–1935) was a Romanian naturalist, glaciologist, and polar researcher.

ch. ends p. 61

was not penetrated, its estimated length of five thousand miles indicated endless land continuity if there had been proper interpretation of the land's existence. And though the dreamer who charted the course to that land was available as the most competent interpreter, his unmistaken interpretation of values was ignored. Thus, no attempt was made to influence a change of popular concept as dictated by the reality then disclosed. For the reality of that land beyond the South Pole holds eloquent refutation of the Copernican Theory's mathematical limitations of the Earth. It was manifest that figures and limitations of theory dominated as arbiters of cosmic reality. Inasmuch as the land's existence and extent did not conform to the established figurative pattern which contributed to popular misconception, its reality had to be denied.

It is easy to grant to a dreamer, who had toiled to have proof established, the right to believe that the proof would actuate questioning of the archaic theory and concept. Perhaps there was such questioning, unknown to him. How much underlying and unexpressed interest that land beyond the South Pole may have aroused can only be conjectured. But it is certain that the expressions of that time could not be considered a token of spirited awakening by arbiters of the cosmic pattern. However, the sensational research and explorative enterprise from 1928 until 1956, undertaken almost exclusively by the U.S. Navy's technical division, attests to a very definite and surprisingly active interest to determine the facts. Yet the reluctance to express interest openly prevailed until a very recent date.

In a final analysis it may be well that organized science, as a medium through which discovered values are interpreted, must adhere to a more rigid procedure than he whose "unnatural" perception enables him to see beyond the acceptable deductive pattern. He who surpasses the pattern owes allegiance only to his soul. It was such quality which permitted discovery of values beyond the ordered pattern. Such being the case, adequate allowances should be made by both sides so that better understanding of the acquisition of values may be had. The lesson should by now be learned that the new and the revolutionary cannot be found in orderly deductive pursuits. Where the extraordinary perceptionist, the inventor, the explorer, or even the creative artist, may and must jump headlong without waiting for the sanction and benediction of tradition's establishments, he must have patience to bide his time until orderly science explores to its own satisfaction the merit of extraordinary findings in whatever field of research, invention, or discovery. On the other hand, it behooves established science to withhold too-ready condemnation of the new and the revolutionary until proper investigation has been made of the new presentation, of whatever nature. There is no excuse for organized science to become impatient.

Accordingly, in the overall word portrait of perceptional values here, it appears to be timely to elaborate upon pertinent features of the fallacious "globe" Earth concept, particularly in relation to the so-called Poles. Some of it may be repetitious. If so, repetition is in order and needs no further apology. This is not a theme so oft repeated of love, hatred, or the many expressions of other human emotion and behavior. This is an original work which has never been published; hence it is necessary at times to repeat the most important and least understood features for the purpose of clarity.

According to the established globe Earth symbol, it must be assumed that any progress beyond the northern or southern geographic centers designated by the Poles would demand a return toward the North Temperate Zone or the South Temperate Zone. The symbol makes such return on the other side a physical necessity. Otherwise — and as the Londoners counseled Sir Hubert Wilkins — one would experience a sharp takeoff into space.

ch. ends p. 61

The misconception of such return from the other side of the globe symbol is so firmly fixed that popular belief holds that the Earth has in fact been circumnavigated north and south on numerous occasions. The belief has persisted despite the fact that there has never been a latitudinal[4] circumnavigation of the terrestrial area. There has been none because there can be none.

It may be claimed that Admiral Peary,[5] Raoul Amundsen,[6] and other explorers "went over the Pole." However, it must also be known that such "over the Pole" accounts have mistakenly represented the term. Its realistic purpose was to show only that explorers *did in fact reach the true Pole points.* To the Poles with a turnabout for return to starting points is possible of accomplishment. But movement to either Pole and "over the Pole" with return to starting point, without turning around, never was and never can be accomplished. It should be realized that explorers of the past did, in certain instances, reach the Pole points. But it should also be realized that they very definitely did not go beyond either Pole and return to their starting point from the opposite side, as popular misconception has held. To and over the Pole point means only movement to and over the assumed mathematical end of the globe symbol, which represents no more than supposed terrestrial extent, whereas over the Pole with continuing movement north from the North Pole or south from the South Pole with return to other known areas of the Earth is impossible.

---

[4] Relating to latitudes north or south.

[5] Robert Edwin Peary Sr. (1856–1920) was an American explorer and officer in the United States Navy who made several expeditions to the Arctic in the late 19th and early 20th centuries, and was long credited as being the discoverer of the geographic North Pole in April 1909, although it is now considered unlikely that he actually reached the Pole.

[6] Roald Engelbregt Gravning Amundsen (1872–1928) was a Norwegian explorer of polar regions, and a key figure of the period known as the Heroic Age of Antarctic Exploration for having been the first to reach the South Pole in 1911, and the North Pole in 1926.

When one goes beyond the Poles one is moving, as the colloquial aptly describes, "out of this world." One then continues to move over land extending beyond the Earth. That land beyond is not on either side of the Earth that was conjectured by Mr. Copernicus. Such a land factor, strange as it may seem to many, is now firmly established by U.S. naval exploration beyond the Poles.

It would be most fanciful to contend that any unknown land existed beyond the Pole points if one believed that the phrase "over the Pole" really means that explorers of the past went over the Pole points from one side to the other side of a supposedly "isolated globe" Earth. Under such circumstances there could be no "beyond" other than the space originally conjectured. But such performance from one side to the other side of an "isolated globe" Earth is an aspect of popular misconception.

The 1928 polar expeditions of Captain Sir George Hubert Wilkins and Rear Admiral Richard Evelyn Byrd, U.S.N., did penetrate beyond the South Pole point in a *southerly direction* and discovered that land extended at least five thousand miles *BEYOND* the original mathematized southern "end" of the Earth. (Incidentally, that estimated five-thousand-mile extent represents the greatest estimate possible through triangulation. And there is no other means for estimating.) Modern expeditions have penetrated into that five-thousand-mile land extent, but its end has not yet been reached. When the end of the estimate is reached, another similar estimate will be made. Such estimating, and penetration to the limit of the estimate, can continue *ad infinitum*. There is no physical end to the Earth, north or south.

That 1928 primary estimate indicated land that continues due south from and beyond what had been considered an "isolated globe" Earth. That land extent cannot be shown by the popular "globe" Earth symbol: it is beyond the bounds of that symbol of theory. But it can be visualized by simply

ch. ends p. 61

adding another globe symbol on top of the South Pole point. The United States and other governments now have land bases on land which cannot be shown by the globe symbol of 1543.

That land beyond the South Pole was seen through extra-sensory perception before human eyes had beheld it and before any mind had deduced its existence. And its reality belatedly established the inadequacy of the four-hundred-year-old conjecture of earth ends and the Earth's relation to the universe about us. The difficulty of average concept to grasp the fact of such physical Continuity of the terrestrial with the celestial has resulted from the fixation that the classroom sphere, depicting the Earth, is a proved entity of the Universe. Such was never the case; it was only a symbol of unproved theory.

The theory of 1543 is extremely abstract. It was evolved by the most abstract science. And its framework, as described here, was based on the inescapable error of lens functioning. No amount of observation, and no amount of increased lens power for magnification of luminous celestial areas, can overcome the illusions developed from such lens error.

Therefore, in the light of values now established beyond the Pole points, one may rightfully question how any physical attempt could have been made to verify the mathematized Earth "ends" when the theory containing such ends was developed. At that time, and until very recent years, there existed no physical means whereby progress could be made beyond the assumed ends for determination that such points were not the ends.

A mathematical designation of earth ends north and south was sufficient for the time of theory. But one should be alert to differentiate between figurative and realistic values of the Universe. By no means is the figure interchangeable with the fact. A famous physicist once referred to that differentiation as follows: "the world of the mathematician is peopled by all sorts of entities that never did, or never could, exist on land or sea or in the universe about us." The apt reference is to the astro-mathematician, whose mathematics ordain a Universe opposed to creative reality.[7]

With understanding that the ancient attempt to interpret the Earth's north and south extent was purely mathematical, it becomes reasonable to question the ends designated by mathematics. Then one can concede the prospect of land and waterways continuing beyond the Poles. With realization of modern discovery which affirms the existence of land beyond, it becomes reasonable to question that land's purpose and where it leads. Then, with acquisition of the observational principles that are firmly established by the sciences, it will not seem out of place to apply such principles in telescopic observation of the universe about us.

The relative relationship of "up" is by no means an innovation by this writer. It has always been known, in spite of the fact that the understanding has not always been afforded practical application. "Up" is always relative to the position we hold anywhere in the Universe structure. When we stand on the land "up there," this terrestrial land we have left behind will have to appear to be "up" to our observation from a celestial area. The fly standing on the ceiling or the floor is as much "up" from either position. Nor is the fly "upside down" when standing on the ceiling. Our concept of values may consider the fly on the ceiling to be upside down, but it can in no way affect the fly's position. The fly stands as firmly on the ceiling as on the floor.

Sitting in the nose of a rocket that is gliding through the stratosphere at an altitude of five hundred miles from the Earth's surface, we will have lost sight of where we entered the dark stratosphere. Then, wherever we look we will

[7] Paraphrased from "Intelligence and Mathematics" by mathematics and philosophy professor Harold Chapman Brown (1879–1943), published in the 1917 book *Creative Intelligence; Essays in the Pragmatic Attitude* (p. 142).

ch. ends p. 61

observe the luminous points astronomically designated. Now, this is the all-important feature very recently proved: as we look toward the sky area covering the land surface we departed from, there will be seen the same luminous points that envelop us from every angle of observation. Then, as altitude is increased, the lights of the celestial will bear no greater relation of "up" than the lights of the terrestrial sky areas. And as the universal sky-light will not be arranged in a direct course over and under our rocket but will appear at every angle, "up" will be everywhere to our observation. "Up" is in fact everywhere. The so-called "Heavens above" are everywhere.

The problem of rationalizing endless land extending beyond Pole points, with the orthodox "globe" Earth concept, precluding any possibility of such land, is conveniently met in the following manner. Grant the imaginary mathematical Poles the physical reality of popular misconception. Let them remain as ends for the Earth of 1543. Continue the Pole points of 1543 to the distance beyond that has to date been penetrated. Mark such points the New South Pole and New North Pole. Then repeat the performance with every exploratory advance made beyond the New Pole points.

As the 1928 explorers beyond the South Pole estimated a land extent of five thousand miles out of bounds of the Copernican "globe" Earth, the extreme limit of that estimate must be considered our New South Pole, when it has been reached. When future expeditions arrive at that New South Pole five thousand miles beyond the original South Pole, they will estimate another five thousand miles beyond the New South Pole.

That Pole-moving procedure will continue as long as men inhabit the Earth and answer the urge to explore such land highways extending beyond both Pole points. And as they continue to penetrate the northern and southern land extensions of the traditional Earth area, they will establish that penetration is being accomplished into celestial areas which, from our present positions on terrestrial level, must appear to be "up," or out.

One may for the present continue to retain the concept of Earth isolation if it is beyond one's ability to relinquish it. The natural course of events will conveniently modify yesteryear's concept without knowledge of the individual. Truth has a very subtle way of entering where it is not wanted. As each successive exploit of man along the northern and southern land highways unifying the terrestrial and the celestial bears confirmation that the Earth is not isolate, the dominant misconception will be dispelled. Such discernment will not come like a sharp hypodermic injection. It will develop like the slow but certain change in growth of body tissue. Then will the Poles of yesteryear's understanding be stripped of their restrictive domination.

It must become most obvious that there are no northern or southern limits to the Earth after explorers have penetrated ten, twenty, and fifty thousand miles beyond the originally assumed ends. And the continuing land being penetrated must therefore represent areas of the celestial. After such extensive penetration, the question would naturally arise: What else can it possibly represent?

Without the stimulus of this perceptional portrait of cosmic values, there has been periodic effort to penetrate the immediate Antarctic Continent this side of the South Pole since the year 1739. However, early explorers were compelled to retrace their course after reaching various points of the vast Antarctic Plateau. They were denied access to the Pole point because of lack of essential mechanical equipment now at hand. And since they could not reach the Pole, they certainly could not have hoped to penetrate beyond the Pole.

The general misunderstanding of southern polar conditions may be realized from the following descriptive account of the Antarctic Continent which bars the course to

ch. ends p. 61

and beyond the South Pole: "A realm of mystery! The Pole is located upon a plateau ten thousand feet high in the center of a vast continent of five million square miles, fifty percent larger than the Unites States. Upon all but one hundred square miles of Antarctica lies a cap of thick ice glittering upon high plateaus and lofty mountain ranges which give the continent an average height of sixty-five hundred feet, or twice the height of Asia."

In the light of modern knowledge concerning southern polar terrain and that area's width, it becomes important to re-examine the four-hundred-year-old concept as it relates to the final quadrant, south, of a supposed isolated sphere. In harmony with the conceptional value originally expressed, can such vast land area and its mountains be explained? In any attempt to harmonize today's discovered reality with yesterday's theory, one must bear in mind that no stretch of the imagination can transform land and mountains into ice.

Recall the elementary provisions of the Copernican Theory that, because of the daily and yearly movements of the supposedly isolated globe Earth on its imaginary axis, the two extremes of that inconceivably rapidly moving globe, or sphere, would accomplish the least movement in time and space. And they would receive less of the Sun's heat as a result of the mathematically prescribed tilt of the Earth "planet" as it made daily movement in its assumed orbit to achieve day and night, while making a secondary movement toward and then away from the Sun to arrange the seasons we experience.

Early interpretation of theory's values held that there would have to be experienced a perceptible tapering of the Earth "body" from the greatest equatorial width to that of the Pole points. However, experience teaches that such condition does not hold. The tapering is imperceptible; it is negative in comparison with the Earth's greatest width. Moreover, in precise conformance to theory, the prescribed movement of theory would demand that the so-called ends be of ice, which is somewhat different from the solid land and mountains found to exist and to be coated with ice. The factor of ice covering for polar areas of the terrestrial results from the position of such areas in relation to the Universe whole, and from the distribution of magnetic force throughout the Universe whole. The magnetic dispensation does vary throughout terrestrial areas in accordance with the natural laws governing its universal distribution.

But the magnetic force of the Creation is by no means dependent on misconceived man-made rules of behavior. Man may assume the structure of the Universe as he will. And he may ordain a fantasy of movement for the continuous Universe structure which his deduction has dissected into multiple disconnected areas. However, and strange as it may seem to man's egotism, cosmic reality makes no provision for man's hopeful but vacuous[8] deduction.

Descriptive material dealing with Antarctica mentions that penguins and whales abound in this previously assumed desolate area of ice and glaciers and "eternal darkness," and that the mountains hold a fabulous fortune in coal and ores. Now reconsider that ancient theory, which to account plausibly for the experienced long days, short days, and seasons as the assumed isolated globe Earth prescribe its assumed yearly course toward and away from the Sun, made it imperative that the assumed ends of an assumed globe would have to be ice. They could never contain the land and minerals of modern record, and the profusion of animal life known to exist.

The awesome decree of the Koran described the northern and southern assumed extremities of an Earth then believed to be flat as "the lands of Eternal Darkness." Are they? The

[8] Mindless or lacking intelligence.

ch. ends next p.

unknown is always fearful and forbidding. Hence it must be considered dark.

As land, mountains, minerals, and profusion of animal life are found to constitute the Antarctic area this side of the South Pole, land, vegetation, and life are to be found as progress is made beyond the Pole and out of terrestrial boundaries.

At that particular Pole point, and for a distance beyond, are experienced the most intense winds and blizzards, which act as a barrier to progress beyond the Earth. Such conditions seem to be an expression of Divine Will which demands that terrestrial man be receptive to cosmic values before he is permitted to penetrate the ice barrier between the terrestrial and the celestial. Beyond the barrier will be found a warmer climate, with land and waterways. And it is there that celestial cousins await terrestrial man's arrival. And if one asks how far beyond, it will suffice to record that the distance is negligible, with modern transportation speed. The northern and southern terrestrial extensions have until very recently been denied in the same manner that the eastern and western water extensions were denied prior to the fifteenth century. Yesteryear's archaic Ptolemaic Theory prohibited terrestrial width because the sky seemed to meet the water at the eastern and western horizons. And the globe symbol, also founded on illusion, has restricted movement beyond the globe's assumed ends. The fifteenth-century experience taught that "things are not what they seem." We have learned that we need not "shoot up" or "shoot down" in movement from one side to the other of an assumed globe Earth. We have learned that we can make such movement without "falling over the edge" of the Earth. Unfortunately, we have not yet collectively learned that we may move straight ahead from the Earth's assumed ends to reach areas of the universe about us which appear "up," or out, from terrestrial position.

The Earth globe symbol would seem to require an up-and-down movement from Boston to Hong Kong, and vice versa. But experience has taught that movement between such points is on the same physical plane. Regardless of what the globe symbol depicts, it should be understood that the Earth's realistic arrangement in the space of its construction is as if both sides of the earth were shown as flat surface areas.

Please don't get lost. This has nothing whatever to do with the archaic flat-Earth concept of the Ptolemaic kings. If one cuts the map surface of the globe symbol from Pole to Pole, and stretches out both sides of the map, it will show the realistic course of movement from Boston to Hong Kong. There is no movement up or down. But the globe symbol must make it appear that there is.

The relation of the entire terrestrial area with the celestial is the same. "Up" is always relative. And we move straight ahead from assumed terrestrial ends to reach the celestial areas which are apparently "up." Or out, from the terrestrial.

Figure 3
This is not intended to show distance from the terrestrial to the celestial; it cannot be drawn to scale. But it does indicate what the nightly view of our terrestrial sky must be for our celestial cousins. Our luminous outer sky, deceptively appearing as millions of rounded and isolated "bodies," would present to the Martian and all other inhabitants of the Universe the identical so-called "Heavens above" which we see as their luminous and deceptively isolated sky areas. Since "up" is always relative, our celestial cousins look up, or out, through their inner blue sky, as we do through ours, and behold the same nightly "star" pattern that we witness.

Contrary to popular misconception based on the illusory, shooting up or out from any location on the terrestrial and the celestial would take the hapless explorers away from the Universe structure and project him into infinite space. Place your thumb on the illustration's stratosphere section, then draw it toward you. That will describe where the space explorer would go, if he did not land back on some land area of the terrestrial. He would be completely lost in space wherein the Universe was constructed, or he would be projected upon some terrestrial area remote from the point of flight origin. Thus the heralded spaceships would be precisely that and nothing else: any spaceship launched (and there is no doubt that it could be launched) would either be lost in space infinite or be returned to some area of the Earth.

Increase of speed and power would hasten the development whereby it would become lost outside the Universe whole. Such is the inevitable destiny for spaceships. The Universe is so ordered that power-increase to overcome the arc of flight would precipitate the spaceship away from the Universe. On the other hand, insufficient power would restrict the spaceship to the movement of all projectiles, and it would have to conform to the arc of flight which would return it to some land area of the terrestrial. That flight principle, always demanding consideration in the firing of our most powerful naval weapons, holds application to the U.S. Navy's superpowered rockets. Their arc and drift is increased with every increase of altitude.

Continuing the study of the illustration for better understanding of the terrestrial "Heavens above," imagine that the luminous terrestrial sky-curves each cover a land area one hundred miles in length and width. Then "cover" the entire terrestrial land with one-hundred-mile sky disks. That will give some idea of the countless luminous "rounded and isolated bodies" our connected and continuous outer sky presents to celestial observation. The results of observation from the celestial would compare with results of our observation from the terrestrial. The magnitude of the terrestrial "heavenly bodies" detected would depend in part on the power of the detecting lenses.

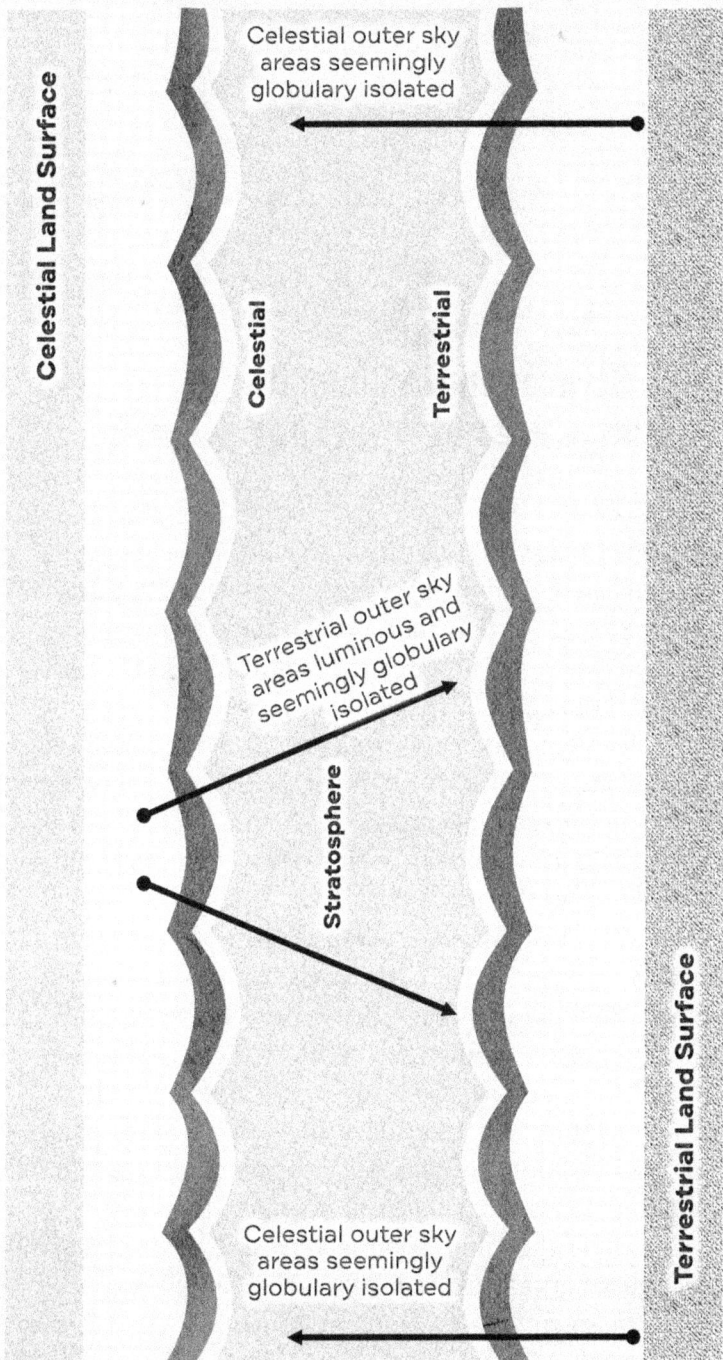

Figure 3

The infinite sky enveloping the universe, showing the inescapable illusions.

Celestial Land Surface

Celestial outer sky areas seemingly globulary isolated

Celestial

Terrestrial

Terrestrial outer sky areas luminous and seemingly globulary isolated

Stratosphere

Celestial outer sky areas seemingly globulary isolated

Terrestrial Land Surface

5
Stratosphere Revelations
"Things are not what they seem."

The pilgrim of 1928 was aware that land discovered beyond the South Pole point confirmed only one aspect of Physical Continuity. He knew that there would have to be photographic confirmation of his disclosure concerning terrestrial sky-light and the deceptively globular and isolated appearance of outer sky areas. Only through such proof could he hope to establish the illusory nature of astronomical conclusions dealing with celestial areas.

Hence his pilgrimage was directed toward procuring the required photographic proof through a stratosphere ascent which would permit photographing an area of the Earth's luminous outer sky surface from stratosphere darkness. Though there had never been a record of terrestrial sky-light, he knew the condition would be confirmed if it was possible for him to ascend into the stratosphere. The lens deceptions contingent upon telescopic observation and photography of luminous celestial areas was most clear to him, but duty to his cause seemed to demand that he spare no effort to show the comparisons at terrestrial level so that others might comprehend the illusions. Therefore, from 1929 until 1935 he sought means whereby he might ascend into the stratosphere. And during that period he recorded the conditions of lights and their movements which produced illusion in the workaday world at terrestrial level.

He relentlessly pursued the mathematical contradictions of theory which had over a period of four hundred years made an incomprehensible patchwork of the universe about us. Though the abstract mathematical values were understandingly applicable in the fifteenth century, when only the abstract could apply in an interpretation of cosmic values, they loomed as poor makeshift in the light of modern research and discovery. For nights without number he patiently observed the brilliant but deceptive beacons of the celestial sky from vantage points on the desert sand and from lofty mountain ledges. In such application he was able to compare the movement of lights observed at every angle on terrestrial level with the seeming movement of lights at celestial level. And he discerned the synonymity[1] of illusions developed from light manifestations at both levels.

The simplest observations held a meaning most profound. And he who dutifully sought the meaning watched and recorded the apparent movement, or "twinkling," of stationary streetlights in Oakland, California. That observation was made from the deck of a ferry plying[2] the seven miles of water from San Francisco to Oakland. Such simple observation proved that the streetlights' *seeming motion* was attributable to the motion of water between his sensitive optic lenses and the lights of Oakland. And it was thereby discerned that known and unknown conditions existing between a telescope lens and luminous gaseous sky areas of the celestial produce the same illusion of motion.

He never tired of experimenting with the play of electricity in the filament of light bulbs of every size and variety.

[1] In this context: sameness; similarity.
[2] In this context, plying is a nautical term that means to traverse a route or course regularly.

ch. ends p. 88

He observed the light's movement from every angle, and under every condition. And such enterprise afforded proof of the influence all light exerts on the optic lens, and on every other lens, for all of which the human lens has provided the pattern.

Observation of the light distortions resulting from magnification of light at various distances provided foundation for understanding of the observational error leading to the absurd astronomical conclusion of "planetary rings." His perception reduced the so-called celestial "rings" to unreal whirling companions of correspondingly unreal astro-mathematico-globular entities assumed to constitute the Universe.[3]

His persistent application and study of the most humble but realistic manifestations at terrestrial level brought discernment of the complete lack of meaning in seeming manifestations at celestial level. The astronomically prescribed celestial features of "puffs of smoke in a barrel," "double stars," "galaxies," etc., were reduced to simpler but realistic values of cosmic expression adequately described in following pages.

The uninvolved play of searchlights on a darkened sky, or other dark area, proved the inability of the lens to record any area faithfully. As the searchlight disclosed that it was compelled to reproduce its circular lens outline on formations of every nature other than globular, it was made manifest that areas not globular in reality were made deceptively globular by the lens.

The distorting influence of mist and fog on luminous areas and objects of the land and the waters contributed to his elaborate ritual of the years. And the study of such influence at work brought confirmation of Physical Continuity before the first photograph of terrestrial sky-light distortion

[3] "Mathematico" is a Latin combining form, meaning mathematical.

existed. And that single feature materially contributed to the premise that the Universe as astronomically assumed to be can never exist.

It was found that halos and rings, and spheroidal intruders of reality's magnificent scene, are found wherever and whenever one seeks them under conditions making for their illusive development. In consideration of the ease with which they are promiscuously[4] manufactured, there is little wonder that they are observed in telescopic observations of the celestial.

He diligently watched and studied the movements of airplane lights reflected against the darkened sky and against the background of other lights in nearby hills and distant mountains. And he was permitted to discern the gross deception the moving airplane lights would impose on the immature mind of some native from an undeveloped region of our civilization. Such a native, lacking knowledge of the altitudinal relation of hills, mountains, and the moving airplane lights and their relation to other lights in hills and mountains and of the celestial sky, would be unmistakably awed by the indefinable spectacle. It was found reasonable to conclude the native's ignorance of the placement and purpose of the various lights, in relation to those of the unknown airplane in motion, would permit no other determination than that the moving airplane lights represented some fearful unknown entity or condition of the so-called "Heavens above."

Though familiarity with moving airplane lights at night enables the more enlightened to comprehend realistic value of the lights and their movement, they are, nevertheless, as readily confused by corresponding light movement and light distortions developed at their immediate terrestrial level. Hence it may be understood that the measure of deception for

[4] In this context: indiscriminately or casually.

ch. ends p. 88

the average person is multiplied by the seeming movement of known and unknown lights at celestial level. Early experimentation established that illusion can readily be fostered in the most astute minds through land surface observation of the light aura which, under conditions favorable to its development, enshrouds[5] an airplane's lights as well as the plane and produces the illusion of a luminous disk moving through the night sky. Inasmuch as a saucer is a disk, the illusion of "flying saucers" is imposed.

It was also proved that haze, fog, clouds, and angles of observation contribute to the foregoing and numerous other illusions. It was further established that even on a very clear night the lights of an airplane in motion present nothing but a "flying saucer" if they are observed through a translucent window glass.

The same illusory developments were found to apply to a bright arclight at the negligible distance of fifty feet from the observing lens as they apply to the "moon" at its estimated distance of about 335,000 miles.[6] And, as distance lends enchantment, the illusion determinable as such at fifty feet is without question accepted as celestial reality when advanced by an astronomical conclusion which holds no possible hope of determination. Though the disguise and projected illusions of lights and luminous areas can be ably penetrated at a distance of fifty feet on terrestrial level, they do, nevertheless, impose temporary deception until investigative determination of their realistic value is had. Hence, consider the enlargement of deception of values.

Observation of the unpretentious flame of an ordinary match eloquently affirmed principles of lens function and deceptions resulting therefrom. Experimentation established that the perpendicular flame of a lighted match in the darkness is automatically distorted by the camera lens,

[5] To envelop so completely that it's hidden from view.
[6] That current estimate seems closer to 238,900 miles (384,472 km).

which, in night photograph, causes the flame to be reduced to a horizontal line. The situation developed in photograph from an airplane at an altitude of only two miles. It was thereby perceived that reducing the perpendicular flame to a flameless horizontal line constitutes primary expression of all lens convergence. An increase of photographing altitude developed the secondary expression in lens function, producing the curve, as previously related. The camera lens curved that same horizontal line up at both ends in the beginning of an arc. On complete lens convergence, achieved at greater photographing altitude, the match presented the photographic appearance of a luminous disk.

The qualification should be made for readers who are unfamiliar with the fact that light is always photographed as white. Hence, though it was known that the white disk represented a luminous disk, the photographed area in a black-and white photograph was white.

This simple match experiment was not considered too simple or unimportant for the United States Army Corps' application of many hours. Therefore, consider what the lens is capable of doing to a straight line and how it can make globular and isolated luminous sky areas that are not globular or isolated. Then it may be possible to reconcile the illusions developing from observation of the celestial with that two-thousand-year-old dictum: "With eyes ye see not, yet believe what ye see not." That parable, too, merits repetition on every page of this book. Its meaning may be generally understood after another two thousand years.

It was found on another occasion that the match flame would, through optic lens function, develop an aura of greenish-red light when held in one's hand and viewed through mildly watering eyes. In other words, there would be formed, by the optic lens detecting the flame through a moisture film, a luminous and colorful circle which seemed to envelop the flame. That illusion in observing a known light not more

ch. ends p. 88

than six or eight inches from the detecting optic lens, and at a time when the least additional moisture between the lens and its object exerted such influence on the optic lens which distorted the object, holds very definite relation to telescopic lens detection of luminous celestial sky areas. Telescopic detection of luminous celestial areas must be had at tremendous distances and through numerous distorting and obscuring media. In some celestial sky-light areas those media become at times much more powerful agents of the illusory than the eye moisture between an optic lens and a known luminous area close at hand.

Though there need not prevail at celestial level a corresponding volume of moisture influencing illusory lens creations seen in the lighted-match aura, there is unmistakable radiation from the gaseous content of all observed luminous celestial areas. The influence of such radiation between the detecting telescopic lens and a luminous celestial area, in conjunction with other conditions of the stratosphere can be expected to develop corresponding match-flame illusion of one and even more luminous circles. Such circles, or so-called "satellites," can then deceptively appear to be circling around the observed luminous celestial area.

At this point it should be explained that it is not only the distorting influence of media through which light is observed, and the function of light itself at the point of observation, which contribute to production of the illusory. There exists beyond such factors the influence which the observed light exerts on the detecting lens. There is expressed the value of "the more you look, the less you see." Too much looking distorts color. Too intent observation of light and luminous areas produces the distortion of light, shadows, or shading. Continued observation of too-intense light causes the luminous area to become *black*.

"Let there be light." Yet the world of illusion is cluttered with light emanations. The Sun becomes a positive bevy of multicolored globes when observed at the angle proper for their development. And in the multiple globes there are multiple smaller globular patterns. The Universe of illusion has no end of globes and spheres and whirling globular "bodies," though none exist in fact.

The terrestrial parallel of heat radiation's power to distort luminous areas and objects was found in observation of a series of wall lights that were clear glass electric light bulbs. They extended at intervals of ten feet along the interior wall of a room one hundred feet in length. The room was heated from open ventilation on the opposite wall ten feet away. From a position on the ventilator side of the room, observation was made of the electric lights at the further end of the room, fifty to one hundred feet away. Hence the heat waves from the open ventilators were between the observing sensitive optic nerves and the electric lights. The motion of the heat waves, though not detected by the optic lens, produced the optical illusion that every light was flickering, or "twinkling." A shift of position to the opposite side of the room, where the lights were seen *without heat-wave interference*, at once permitted observation of the realistic unflickering lights, thereby proving the illusion.

It is significant to note that this illusory condition was found to develop when the heat waves lacked sufficient force and volume *to be seen by the optic lens*. The radiation exerted its illusory action though it was not seen as a barrier to and distorter of light observation.

Earlier a counterpart of heat waves' influence was shown in the influence of water motion on the sensitive optic nerves as the optic lens detected streetlights in Oakland. Under such conditions of observation, the larger and more luminous streetlights were subjected to corresponding influence, and they afforded the same illusory performance. However, it is pertinent to record that the streetlights' movement was *more pronounced at a distance of five to seven miles* than the

ch. ends p. 88

illusory movement of electric lights at distances of from fifty to one hundred feet.

There is a lesson here of greater illusory movement with an increase of distance from observed luminous area. It has considerable to do with the Galilean premise of illusion, "rounded bodies circling or ellipsing in space." Consideration of astronomical distances should bring understanding of Physical Continuity. And it should assist one to know that movement may be had from the terrestrial Poles into the universe about us.

As this is written, a tiny voice seems to bring an astronomer's expostulation[7] that no such deceptions can be imposed upon the magnificent lenses of astronomy's workshop. And it contends that the greater power of telescope lenses penetrates the conditions that create the illusory. Therefore, it should be said that no amount of light magnification can produce greater clarity. The light and the lens seem to resent magnifying: Increased magnification of light and luminous areas develops a greater volume of light distortion. It becomes evident that the brilliant writer of yesteryear, Tiffany Thayer,[8] was cognizant of such a feature when he referred to the two-hundred-inch telescope lens then being perfected as "the white elephant of Mount Palomar."[9] That lens is competent to *magnify all the illusions of the centuries.* Lens magnification of light and luminous areas, and the light distortion that ensues, is that which produces "canyons" on the Moon and a grotesque array of astronomical entities "that never did and never could exist on land or sea or in the universe about us."

---

[7] An objection or protest.

[8] Tiffany Ellsworth Thayer (1902–1959) was an American actor, writer, and one of the founding members of the Fortean Society, a New York City-based society founded in 1931 to promote the occult, supernatural, paranormal, and anomalous phenomena ideas of American writer and researcher Charles Hoy Fort (1874–1932).

[9] Thayer, T. (1942, January). Circus Day is Over. *The Fortean Society Magazine* No.6 (p.2).

Light magnification is the imponderable which produces the *light shadings* in luminous celestial areas. Such light shadings within luminous sky areas are at times heralded as "clouds" in the stratosphere over the celestial sky-light area; at other times, they are claimed to be vegetation on the celestial land under the sky-light.

At this point it is well to repeat that telescope lenses cannot penetrate celestial sky-light. It is true that clouds and vegetation are helpful to human beings. Without the clouds vegetation might not exist. Hence one may take one's choice as to what light shadings represent, other than light shadings. Though clouds and vegetation exist under the light which extends throughout the Universe whole, such conditions cannot be detected through the luminous sky envelope. All that telescope lenses detect is an aspect of the luminous sky.

These and innumerable corresponding truths of experimentation and brain observation have been developed through unremitting effort to refute or to verify the disturbing perceptional portrait of the realistic Universe. For that portrait was presented to that early pilgrim as a burdensome and heartbreaking gift from the Force which ordains out individual destinies. The gift could not be rejected, because the Force persisted in its endowment. But is it to be wondered that he who was so endowed made periodic attempts to abandon the gift? The hours he consumed in tedious combing through the centuries accumulation of astro-mathematical data embodying glaring contradictions that resulted from organized endeavor to sustain the postulate of terrestrial isolation constituted a period which could have thrice told the fables of "a thousand and one nights" fame.[10] And time would have been left to erect all the unreal mathematical universes that history records.

To accomplish a project of such magnitude that it opened

[10] *One Thousand and One Nights* is a collection of Middle Eastern folk tales compiled in Arabic during the Islamic Golden Age.

ch. ends p. 88

the centuries ice-blocked paths to the universe about us, that early pilgrim's elaborate laboratory was generally the uncluttered platform of the desert sands. And his customary astronomical observatory was an unsheltered mountain ledge. But his equipment was superior to the most powerful telescopes of Mount Wilson and Mount Palomar. At the latter, the two-hundred-inch lens was then being ground and primed "to see all and know all." *Absurdum! Absurdum!*[11] It is the brain that truly sees. And telescope lenses do not have brains.

His endowment fund was of flawless extrasensory perception, which had detected more of the realistic Universe in five minutes than all the telescope lenses of the ages could detect. And his loyal organization was faith — his faith against a world of skepticism.

In 1932 he met the Belgian stratosphere explorer, Professor Auguste Piccard, at the professor's quarters in the Hotel St. Moritz in New York City.[12] It was there that he viewed the first photographs of the terrestrial outer sky that he had described before any lens had detected it. Piccard's photographs showed a minute area of the Earth's sky as it is to be seen and photographed *from within the sky*. The photographs had been taken at Piccard's greatest altitude, and that was only on the threshold of the stratosphere. Piccard had not achieved sufficient altitude for a photograph against the stratosphere background of total darkness. Hence the photographic plates showed only the lower sky area through which Piccard had entered. That sky area appeared as "an illuminated upturned disk."[13] The corners of that upturned

[11] Translated from Latin: Absurd! Absurd!

[12] Built in 1930, on the site of the old New York Athletic Club, the Hotel St. Moritz was a luxury hotel located in Midtown Manhattan, and today is known as The Ritz-Carlton New York, Central Park.

[13] The actual quote is: "It seemed a flat disk with upturned edge." As found in: Ten Miles High in an Air-Tight Ball. (1931, August). *Popular Science Monthly* 119(2), p.23.

disk were developing a copper tinge representing primary illumination of the immediate sky area. It was the color seen on cloud formation as the Sun disappears far beyond the western horizon.*

That illuminative coloring of the upturned or partial disk obscured the outline of the terrain where Piccard's ascent originated. Nothing of the Earth's surface was to be detected by Piccard or by the camera lens in the base of the stratosphere gondola. All that could be seen was the partly luminous partial-disk development of the sky area being penetrated.

Though Piccard had not achieved sufficient altitude to permit the lens formation of a complete disk with total luminosity, his photographs confirmed lens function and the resulting deceptions as disclosed since 1926. If he could have increased his altitude, the partial, or upturned, disk would have been completed *by the lens* into a full disk. Both edges of the upturned disk, as shown at the beginning of stratosphere darkness, would have been continually drawn up by the lens until they met. Then the upturned disk would be detected from stratosphere darkness, and from all other areas of the Universe, as a down-turned curved area. When that condition exists, there is presented a complete disk surface, which is known as a disk. We do not speak of down-curved areas; when they present such formation, they are known as disk-like.

The lens completes the circle because the lens is circular. With completion of the circle, the disk area is detected; the lens has done its job. Then the mind adds the finishing touch, which causes the illusory circular outline of the sky area to have body property. The fullness of body must exist for

* In this analysis it is important that understanding be had of the sky depth. The sky is not just a blue film on one side and a luminous film on the other side. It has a measurable depth. In other words, there is sky density.

ch. ends p. 88

the adult mind, though there be no such fullness of body in reality.

The appearance of that particular sky area being photographed in 1931 impelled Piccard to announce: "The Earth appeared as an illuminated upturned disk." However, it is self-evident that Piccard meant that *the photographed sky area* appeared as an illuminated upturned disk.

The word "illumination" has application in this instance because there was illumination. But there was no luminosity. There was not sufficient darkness of stratosphere background for luminosity to develop. Though the sky area being photographed from within the sky depth was not luminous, the primary illumination was sufficient to obscure the land surface. Only increased altitude, with additional stratosphere darkness, would develop luminosity.

Piccard acknowledged in the early descriptive account that he could see nothing of the land surface: "A copper-colored cloud enveloped the Earth." There is no doubt that Piccard meant well. But he, or the journalist quoting him, used an exceedingly misleading choice of words. As herein related, (1) the Earth did not appear as anything, because no area of the Earth could be seen or photographed. (2) It was only an infinitesimal area of the Earth's entire sky that provided the appearance of an "illuminated upturned disk." (3) The "copper-colored 'cloud'" was part of the gaseous sky density which was developing luminosity.

Accordingly, the photograph plates displayed by Piccard afforded ample evidence that he had not achieved sufficient altitude for lens development of the complete disk. Had Piccard gone beyond the outer sky surface and photographed from stratosphere darkness, the complete disk would have been developed by the camera lens, and the dark stratosphere background would have caused the gaseous sky illumination to posses fiery luminosity.

Observe **Figure** 4 at the end of this chapter [pp. 89-90].

Released to the nation's science editors in 1930, it shows how every area of the Earth's luminous outer sky would appear from sufficient distance in stratosphere darkness and from all celestial land areas. If the luminous disk-like areas were to be drawn into complete circles, the lower half of each would describe the "upturned disk. In viewing any luminous sky area like those shown from the depth of stratosphere darkness and from celestial land areas, the half-circle curves are presented as disks. There should be no confusion on that point. The feature could have been established in ancient Babylon if they had possessed V-2 rockets.

Unfortunately, when the luminous disk-like areas are detected at terrestrial or celestial level, *the human mind automatically provides body property* which does not exist. In such manner does the realistic Universe become infested with "isolated globes" that do not exist. The Earth area of the Universe whole could not escape the "isolated globe" infection. Astronomical dogma decreed that the luminous celestial areas detected were "isolated globular," or spherical, "bodies" adrift in space infinite. And such being the case, the Earth had to be the same. Who could prove it otherwise in 1543, when the theory of "astro bubbles" was imposed?

While we are at it, it might be well to turn to **Figure 5** [at Preface] entitled "The U.S. Navy's V-2 rocket camera photographs dispel the illusion." The title is most fitting. The photograph shows a luminous outer surface area of the Earth's sky from an altitude of sixty-five miles. "Altitude" means distance from the Earth's surface; hence the photograph was taken approximately fifty-five miles beyond the outer sky area. It might have been a little more than fifty-five miles, because the distance from land surface to sky varies at times and at different places the sky is only seven miles away, at other points it may be ten miles.

Figure 5 is a reproduction of the original V-2-rocket-camera photograph of a small area of the Earth's entire sky. The

ch. ends p. 88

photograph was not taken on the perpendicular, as was the case in Piccard's photograph of 1931. Hence it shows only *at an angle the complete disk area* which Piccard's perpendicular photography would have shown if he had ascended to the V-2-rocket height. The rocket camera would have shown a round disk, rather than a foreshortened oval, if it had been in the rocket's tail so that it could photograph on the perpendicular during the rocket's ascent. As the rocket descended it was drifting at an angle; hence all photographs of the outer sky had to be at an angle. Had the rocket avoided drifting and descended in a perpendicular course, it would have shown the full disk area indicated in **Figures 3** and **4**.

That original photograph of an area of the Earth's luminous outer sky surface, *seemingly globular and isolated*, is the most important photograph in the history of the world. It tells more of the realistic universe about us than all the astronomical volumes compiled throughout the centuries. It needs only the proper interpretation. And if terrestrial man is not competent to interpret its meaning at this time, he should be denied acquisition of the universe about us.

The white area of the photograph is the luminosity which covers all sky areas. The dark areas depict light shadings developing from the gaseous movement which produces the light. Other factors may have influenced the shading as shown. If it had been a very powerful automatic lens photographing from that distance, magnification of the light surface would have occurred. Then the clear luminosity could be considered to exist only in the white patches. But that conclusion would be faulty; the light covers the entire area. From greater distance it would become manifest.

There was reference earlier to such light shading being "cloud formation." That term is acceptable if it means *gas-cloud formation*. Otherwise it becomes ridiculous. If the shading or the white patches were in fact atmospheric clouds as observed from land surface, the surface of the Earth could also be detected. Nowhere does it appear; and it could be made to appear only through the application of an appropriate photographic medium capable of penetrating light. There is such a light-penetrating medium developed by modern research, but its application can always be detected because the object or area photographed *through* light becomes distorted. As illustration, green vegetation is reproduced white, and the normal contour of objects becomes out of proportion.

Though the medium referred to, infrared and extra-sensitive film, has application to photography within distance limits, there is no record of its application to telescopy.

If there existed, or if there is ever developed, a medium whereby telescopic lenses can penetrate the luminous celestial sky-light, even astronomers will then be permitted to discern the factual universe about us. Then will they observe the land underlying the luminous outer surface of celestial sky areas where astronomical conclusion has denied the existence of land. Then will they detect the abundant water and vegetation denied by astronomical conclusions of the centuries. And that vegetation will give the lie to the astronomical assumption that celestial areas lack the oxygen content conducive to life.

No astronomer, or his most powerful telescope lenses, has ever detected more than the luminous outer sky surface of any area of the universe about us. No telescopic camera ever photographed other than the same sky surface area which is made deceptively disk-like and isolated by the lens function described here. Therefore, strange as it may seem, photographs of luminous celestial areas with fullness, or body, are products of illusion. The tragedy of their display is expressed by the misconception they foster. The lens-formed disk area of celestial sky is the only thing photographed, but the disk area must develop the delusion that a full and isolated body exists.

In view of **Figure 4** showing what every terrestrial sky

ch. ends p. 88

area would appear to be from the dark stratosphere and from other land areas of the Universe, Professor Piccard's photographic development of a partial disk with incomplete luminosity was not generally considered as evidence of the illusions described. Effort was therefore intensified to have photographs of the Earth's luminous outer sky made from greater altitude which would show a complete disk with luminosity. The requisite altitude was considered fourteen miles, four miles beyond Piccard's altitude.

With that objective, a journey was made to the U.S. Army Air Corps' base at Wright Field at Dayton, Ohio. There Major Hoffman[14] and Captain Albert W. Stevens were making elaborate preparations for a stratosphere ascent, and it was believed that they could be induced to achieve a four-teen-mile altitude, where photographic confirmation of lens deceptions would be had.

Captain Stevens, then considered the leading aerial photographer, had taken numerous photographs of the business section of Dayton, Ohio, at an altitude of five miles. Photographs from that altitude, doubtless with a very powerful camera lens, showed the known concrete structures of the business district being *merged together* by lens function. Such merging confirmed that photographs at greater altitude would cause the concrete structures deceptively to appear as rounded or globular.

Although the converging function of all lenses had long been established, the extraordinary photographs reasserted known principles and contributed additional knowledge that lens function can create innumerable illusions at terrestrial level. And the illusions would develop from observation of objects and conditions with which we are most familiar. Hence it was not difficult to determine that there would be

---

[14] Edward Lincoln Hoffman (1884–1970) was a United States Army Air Service pilot, officer, and Engineering Division Chief who co-developed the first modern parachute while stationed in Dayton.

multiplication of the quantity and quality of lens-developed illusions in telescopic and photographic observations of remote luminous celestial areas which are entirely unknown.

Aerial photography has likewise established the gross deceptions resulting from altitudinal photographs of familiar terrestrial terrain where rivers, seemingly drawn to the surface of the land and deprived of natural depth and width, lost their identifying characteristics as rivers and were made deceptively to appear as streaks on the land surface.

Through the courtesy of Major Hoffman and Captain Stevens, the pilgrim of 1934 inspected the stratosphere-ascent equipment at Wright Field and prescribed the altitude required for photographic confirmation of his earlier claim. The minimum altitude considered necessary was fourteen miles: ten miles from land surface to sky and four miles into the stratosphere darkness beyond the sky. Captain Stevens gave assurance he would make every effort to achieve the required altitude. His initial attempt failed when the balloon burst shortly after the ascent was under way. Soon thereafter, November 1935, the ascent attained fourteen-mile altitude over the Black Hills of South Dakota. There is little question that at that altitude were made confirmative photographs showing complete luminosity and disk appearance of the sky area. Unfortunately, the photographs of that ascent were not released when requested.[15]

There was no further important development bearing upon Physical Continuity until October 1946, when the U.S. Navy's V-2 rocket achieved the unprecedented altitude of sixty-five miles. And its camera returned sensational photographs of an angle of a luminous, globular, and isolated sky area over White Sands, New Mexico. More recent rocket-camera photographs from an altitude of two hundred miles (May 1954) show a luminous terrestrial sky area

---

[15] But we Heathens have included one of those photographs. See p. 46.

ch. ends p. 88

estimated at three hundred thousand miles wide. It too is deceptively globular and isolated.

In comparing such rocket-camera photographs (made possible by the U.S. Naval Research Bureau) it is important to observe that the globular and isolated appearance is produced at every photographing distance from the outer sky surface. There are no variations of contour; but there are variations of the light shadings and light distortions, which this work has properly stressed.

With such conclusive photographic evidence of terrestrial sky-light and the lens-developed deceptions of the sky's contour, there was reason to believe that some acknowledgement of the pilgrim's claims would be forthcoming from the established scientific order. It was reasoned that such vivid expression of lens deceptions would enable the most skeptical to perceive that identical deceptions were experienced in all telescopic observation of luminous celestial areas. As a result of the sensational rocket-camera proof of celestial and terrestrial sky-light synonymity and the apparent globularity and isolation of sky areas, it seemed that almost anyone would realize that astronomy has dealt only with celestial sky-light and illusory features developed by telescopic lenses.

The illusions now proved to develop from telescopic observations of the celestial attest that it was natural for Copernicus, Galileo, Newton,[16] and others of their times to conclude that luminous celestial areas are globular and isolated "bodies." The illustrious gentlemen lacked modern mechanical devices for proving otherwise. Such being the case, it was necessary to prescribe mathematical space orbits for the *seeming movement of such illusion-born entities* comprising the Universe. Thereafter, the concept of "body" and

[16] Isaac Newton (1642–1727) was an English author, mathematician, physicist, astronomer, alchemist, theologian, and the culminating figure of the Scientific Revolution of the 17th century.

"bodies" became so fixed that it was impossible to return to the 1543 starting point for investigation of the premise. Though numerous men have questioned the premise, there could be no constructive investigation in the absence of more recent mechanical equipment. It is only through timely development of such equipment that proof has been had of the concept's development from the illusory.

Apropos of that mechanical equipment, the early years of pilgrimage led to the cell-like laboratory of Dr. Robert Goddard, pioneer extraordinary in rocket construction.[17] When in 1926 the pilgrim visited him at Clark University in Worcester, Massachusetts,[18] he desired the rocket's perfection no less ardently than Dr. Goddard did. At that time twenty-eight years old, he did not dream that he would live to witness the rocket's spectacular performance, which has developed confirmation of his dream.

Yet with all of modern mechanics, which surpass the mechanics draped in the corridors of time, it took twenty years to utilize that rocket for proof of most sensational disclosures. And it is significant that such a powerful mechanical instrument for proof was first used to destroy. It might not have been utilized for profound scientific purpose but for the fact that the military had become seriously interested in rocket development to meet the challenge of the worst war in history: World War II.

In the proof now established for this work's principal features exists a parallel to the outlook of 1493, when a "New World" of land and water and life which archaic theory had denied was discovered. This land's existence had been denied

[17] Robert Hutchings Goddard (1882–1945) was an American engineer, professor, physicist, and inventor who is credited with creating and building the world's first liquid-fueled rocket, which was successfully launched on March 16, 1926.

[18] Founded in 1887 with a large endowment from its namesake Jonas Gilman Clark (1815–1900), Clark University is a private research university in Worcester, and was one of the first modern research universities in the United States.

ch. ends p. 88

as a result of an illusory condition accepted as real, the sky meeting the water. To overcome the "flat Earth" concept developed from that illusion, it was of utmost importance for science to make plausible the existence of this "New World." Hence when there was advanced in 1543 a forceful ease to sustain the timely, though erroneous, concept of isolated celestial areas making up the so-called "planetary system" and simultaneously explaining the New World's existence, it was most acceptable.

At that time, it was of primary importance to establish convincingly that the Earth's area, previously conceived to embrace only the Old World, was in fact twice as large. And, to give assurance that one would not "fall over the edge," it had to be shown how journeys could be accomplished from one side to the other side of what was considered a globe Earth. The feature stressed was that of the eastern and western water extensions then recently discovered to connect the Old World with the New. The width had to be known, regardless of what the length might be. Progress was east to west rather than to north and south. If there existed northern and southern extensions, it was unimportant to the time.

Promulgation of the globe-Earth idea was simplified by the evident fact that the Sun rises in the East and sets in the West. And it was further advanced through understanding that one could sail to the West and ultimately arrive at points in the East. It appeared reasonable to assume that the Earth's contour was that of a globe, or sphere. Since the assumed globe Earth had eastern and western limits in time and space, mathematical northern and southern limits which would make it conform to a sphere had to be provided. Thus mathematical formula decreed that Earth contour is comparable to that of assumed globular and isolated celestial areas. Though the assumed globularity and isolation of celestial areas, has since been proved illusory, the frames of theory were obliged to accept such apparent conditions as fact. Hence the terrestrial, also assumed to be a globular and isolated area of the Universe whole, like celestial areas was also assumed to be "circling or ellipsing in space."

The ancient theorists, lacking modern equipment for determination of cosmic reality, were convinced that the telescope lens was a faithful recorder of celestial conditions. Unfortunately, lens capriciousness was never considered in determination concerning arrangement and movement in the Universe. However, from that faulty assumption of globularity and isolation there developed a basis for precise time measurement. Where previously the terrestrial day could be known with but two parts, the periods of light and darkness, the theory of terrestrial globularity and isolation made it possible to gauge the light and the dark periods through the application of hours. And the hours, naturally, corresponded to the assumed Earth sphere's assumed daily movement.

It may be perceived that the same time measurement could have applied if, contrariwise, it had been assumed that the Sun described a daily course around the Earth from east to west. Then it would have made little difference if the earth were assumed to be globular, cylindrical, or tubular in contour. Sun movement could provide hours of the day as readily as Earth movement did.

The assumed circling movement of the assumed Earth sphere was made to conform to the time gauge, and the time gauge conformed to the assumed movement of the assumed earth sphere. Hence the mathematized approximate twenty-four-thousand-mile circumference of an assumed globe Earth invited mathematical determination that one twenty-fourth of the Earth's assumed daily turn in space would constitute one hour. Therefore, since one assumed complete rotation of the assumed globe Earth of twenty-four-thousand-mile circumference would constitute an Earth day of twenty-four hours, there had to be twenty-four different starting points

ch. ends p. 88

for time, every thousand miles of the twenty-four-thousand-mile circumference would factually experience a different twelve o'clock noon and a different twelve o'clock midnight. Such mathematizing was by no means complicated.

It then followed that the diameter of the man-made globe Earth would have to conform to global dimensions. Accordingly, there had to be formulated assumed northern and southern diminishing points for the assumed globe Earth assumed to be isolated in space infinite. Reality could not be consulted, and it could in no way control designation of the assumed northern and southern ends sustaining the globular concept and the isolated Earth globe.

Man, having established the Earth's contour and limits to meet the need of that time, had very little interest in the physical aspects of the northern and southern extremities which his mathematics had ordered. His interest was centered in travel east to west from the "Old World" for conquest of the western "New World."

After the assumed globe Earth's assumed ends were mathematically fixed in time and space, there had to be provided an independent orbit, or space path, for its assumed daily and yearly movement in relation to other assumed cosmic "globes" scattered throughout timeless infinity. They, too, had to be made to conform to the mathematical order perfecting man's illusory Universe.

Hence it may be perceived that man, rather than Creative Force or Deity, was responsible for the fifteenth-century pattern of the Earth and the universe about the Earth. Nevertheless, the pattern woven from illusion served a purpose and filled a need of that time.

It can be readily realized that the interest of four hundred years ago could not, and need not, be in any constructive manner directed toward the assumed ends of the assumed Earth globe. Lack of factual Knowledge of the Earth's northern and southern extent explains why the most famous of American explorers as recently as February, 1947, was impelled to describe the endless land extending beyond the assumed northern end of the Earth as "the center of the great unknown."

Though the Universe structure imposed by the Copernican Theory was developed from illusion, the misinterpretation of values bestowed certain benefits upon men of that era. It afforded adequate general understanding of this "New World" reality. And it provided a necessary and most helpful gauge of time even though, in so doing, it prescribed a series of fanciful movements for assumed cosmic "globe bodies" which, in common with the assumed Earth "globe body," seemingly constitute the Universe whole.

Unfortunately, in providing such benefits there also developed the very questionable benefit of belief that man would "fall off" the Earth ends north and south instead of the Earth's "edges" east and west. Theory may persistently oppose theory, but only fact can displace theory. The facts of our time disclose the fallacy of assumptive Earth ends north and south. Such facts of modern discovery provide abundant evidence that land and water extends indefinitely beyond both assumptive ends prescribed by theory of 1543.

Twenty years of deepening research into stratosphere darkness confirms the 1926 disclosure that every area of the Earth's outer sky surface, regardless of its size, presents a photographic replica of all that which has been observed of the universe about us. This feature alone provides conclusive evidence that "things are not what they seem" throughout the created Universe. It proves that telescopically detected celestial light is the same sky-light that has been proved to cover the Earth.

Hence it is established that underlying all celestial lights is the same atmospheric density as that of the Earth, which makes the sky possible. It is shown that the gaseous sky content making our outer sky surface luminous against dark

ch. ends next p.

stratosphere background is the same gaseous substance making celestial luminosity. The sky and its light prevail even where no telescopic lens detects them. There are certain areas of our terrestrial sky-light that cannot be detected in telescopic observation from land areas of the celestial. But that lack of detection in no way confirms the absence of terrestrial sky and its light. Therefore, in the modern facts of discovery confirming the presence of similar celestial sky-light and atmospheric density as that which is known to prevail at terrestrial level, there is sufficient evidence that similar terrestrial vegetation and life exist throughout the Universe whole.

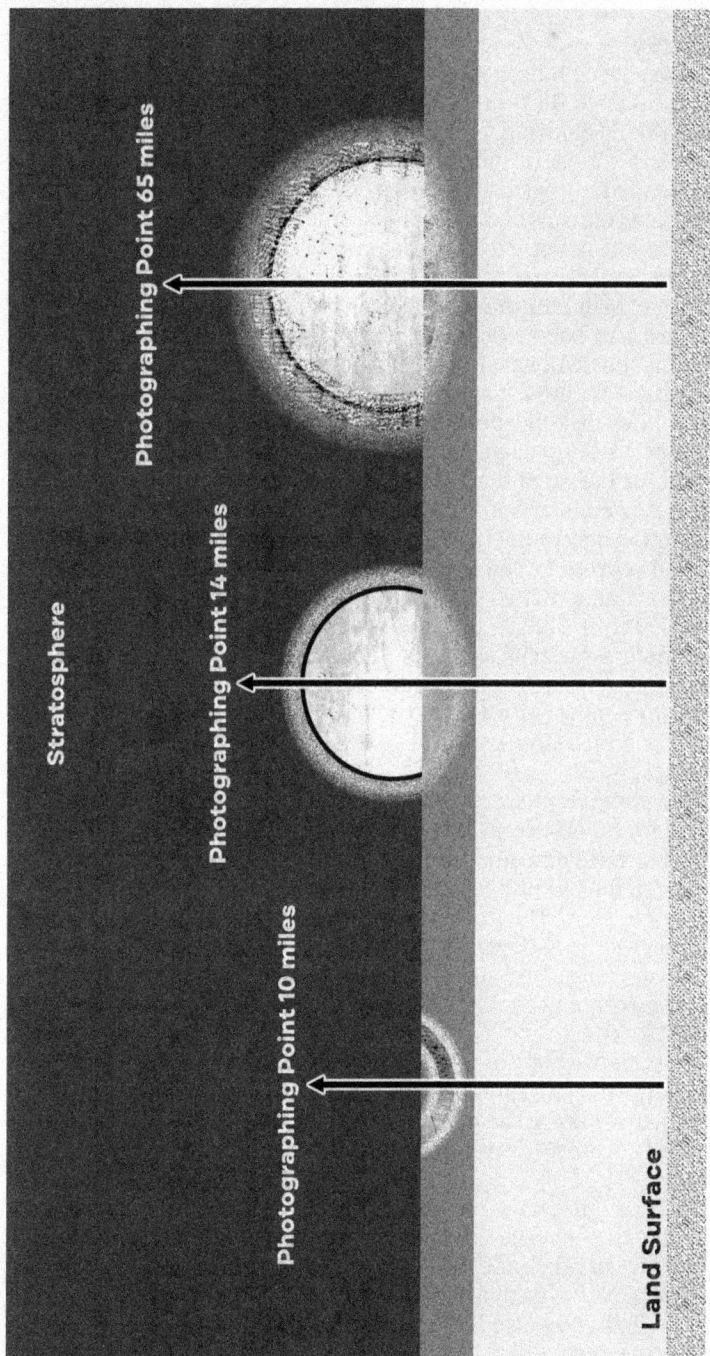

Figure 4

Apprehending the act of deception in stratosphere photography.

Figure 4

Stratosphere photographs prove how the lens develops curves which are seen as disks. They are purely illusory, and they impose the globe-body delusion.

This triple illustration expresses the historical sequence of events confirming the camera-lens development of the deceptive curve. They confirm the Physical Continuity of the Universe.

1. On the left is depicted the beginning of curve development by the camera lens utilized in Auguste Piccard's stratosphere ascent of May 1931 that achieved an altitude of ten miles. Where Piccard had barely penetrated through our familiar blue sky, there is shown the beginning of lens-produced curvature of that particular sky area. It appeared as an illuminated upturned disk.

2. The center disk-like development shows the deceptive appearance of the sky area penetrated by Albert W. Stevens, of the U.S. Army Air Corps, at the greater altitude of fourteen miles over the Black Hills of South Dakota in 1935. The greater altitude permitted development of full curvature, which is detected as a disk. It represents completion of lens function, which develops the "partial upturned disk" into a full disk.

3. The larger and more luminous disk at the right represents a luminous terrestrial sky area photographed by the U.S. Naval Research Bureau's rocket camera at the greater altitude of sixty-five miles, or about fifty-five miles from the sky's outer surface which varies from seven to ten miles from the Earth's surface.

These photographs, and others that followed at altitudes up to two hundred miles, conclusively confirm the disclosures of 1927, that the Martians and other inhabitants of the Universe are obliged to consider that luminous disk-like area over White Sands as a "planet" or a "star." The photographs establish that every Earth sky area observed from beyond the Earth must deceptively appear as an isolated "globe body" comparable to the many luminous celestial areas of astronomy's fallacious "star chart" which is in reality a celestial sky chart.

Camera lenses of the stratosphere ascents and rocket flights were unable to penetrate through the impenetrable luminosity of our immediate sky at the negligible distances involved. Therefore, they could not detect the realistic land and life we know to be under the sky.

Telescope lenses, including the recent two-hundred-inch lens, are unable to penetrate through the luminosity of celestial areas to detect the equally realistic land, vegetation, and other life, existent under every area of celestial light and all other celestial areas where no light is detected.

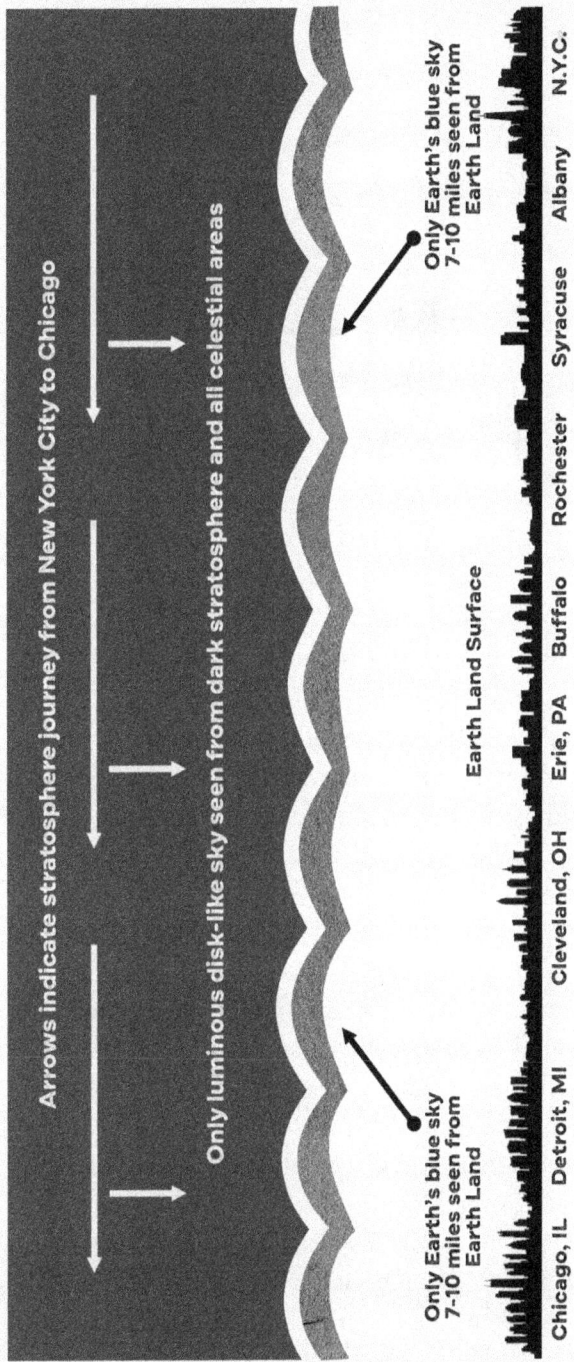

Figure 6

A thousand-mile stratosphere journey over the Earth's sky-light road of illusions.

NOTE: This illustration was originally presented to the science editors of this nation's press services prior to procurement of any stratosphere photographs of our Earth's luminous disk-like-appearing sky segments. The U.S. Naval Research Bureau's V-2 rocket camera photographs, since October 1946, conclusively confirm the presentation.

On a thousand-mile stratosphere journey, from New York City to Chicago, our luminous and illusion-producing outer sky, because of lens-developed curves, deceptively appears as numerous rounded and therefore seemingly isolated "bodies" identical to astronomy's fictional celestial pattern of "stars" and "planets." Though the inner blue sky and the outer luminous sky are both shown to complete the illustration, it must be remembered that the Earth's blue sky is seen only through our Earth's atmosphere, whereas the Earth's luminous sky is seen from stratosphere darkness during day and night and from all other land areas of the Universe during night's darkness.

6

A Journey Over the Earth's
Sky-Light Road of Illusions

The lens is the culprit,
And the deception is the crime.

Figure 6 is reproduced from the 1930 original released to the science editors of press syndicates in New York City. It is intended to show lens deceptions experienced in all observations of the Earth's luminous outer sky surface from stratosphere darkness and from other land areas of the Universe. It was also intended to indicate the lens deceptions resulting from telescopic observation of luminous celestial areas.

Though the drawing was made prior to any confirming photographs of stratosphere ascension or rocket flights, it may now be viewed as reality, because of the V-2-rocket photographic confirmation since October 1946.

1. The land area, as indicated at the bottom of the drawing, represents the accustomed location in our observation of the familiar blue sky between New York City and Chicago. In looking up, or out, from such land positions — or from any other land position of the Earth — we observe the blue

sky of varying depth, or density, from time to time and from place to place.

2. The sharp horizontal curves are never experienced with such sharp angles. The abrupt termination of the horizon is here required to complete the illustration. It imposes lines of demarcation[1] between the various land communities. It also permits simultaneous view of inner and outer sky curvature. The outer are to be observed only from stratosphere darkness and from other land areas of the Universe.

3. The region between represents the seven- to ten-mile distance from land to blue sky. The distance varies over the Earth, and over the Universe whole. Inhabitants of other land areas of the Universe can view no other blue sky than their own. They cannot see our immediate blue sky, but they do see our outer sky surface as we see their outer sky surface. At night, they view our sky's outer surface areas, and every sky area, as here depicted, is luminous and deceptively globular. Hence the deceptive globularity imposes the appearance of isolation. Accordingly, our terrestrial area appears to other inhabitants of the Universe as the same isolated "stars" and "planets" as their areas appear to our observation. Our sky areas make their "Heavens above," as their sky areas make our "Heavens above."

4. The dark area of the illustration above the sky areas represents the stratosphere, which extends indefinitely. As it encroaches upon terrestrial sky areas, it likewise exists over all other sky areas of the Universe.

5. The luminous and disk-like outer sky areas show how the gaseous blue sky of terrestrial land observation becomes luminous against the dark stratosphere. The lens detecting such luminous areas, which we definitely know are not globular and isolated, is compelled by its function to create the

[1] Division.

ch. ends p. 122

curves that produce the luminous disk areas as illustrated. Each disk area must, as previously explained, impose that further illusion of a body. The celestial "bodies" of astronomy are precisely what the illustration describes.

Hence from a distance we see the illustration's luminous disk-like areas as true disk surfaces. Likewise, do we observe luminous celestial sky surface areas, the so-called "stars" and "planets" of astronomical assumption. And inhabitants of celestial land areas view luminous areas of our sky in precisely the same manner as we observe luminous areas of their sky. In sharing our lens illusions, as they must, they have been denied access to the terrestrial in the same manner that we have been deprived of physical journey to their land.

Since the drawing could have no purpose if the complete disks were shown, it portrays only half disks, or a series of luminous arcs. That is all that is really required, inasmuch as that alone is what the most powerful telescopes are able to detect throughout the Universe. If the lower blue-sky areas of the illustration were obscured as one held the illustration at arm's length and observed from the top of the page, one would discern that any area shown would appear as a disk from distant observation. As explained earlier, when that deceptive lens-formed disk area is detected, the mind automatically supplies the fullness which completes the disk and imposes the delusion of a "globe body." Every luminous outer sky area of the Earth and the Universe about the Earth must, *through lens function and only thereby*, be detected as a disk-like area illustratively presented, and it is then *assumed* to be a globe, and the illusory globe must appear to be isolated.

It should be understood that every luminous arc, or disk-like sky area as illustrated, possesses width as well as length. Since there are nine luminous sky areas in the distance, or length, of stratosphere course from New York City to Chicago, each area should be considered approximately one hundred and eleven miles in diameter, to make the approximate thousand miles between New York City and Chicago. It may be considered that in the flight machine photographing that sky course there will be a lens of sufficient power to embrace an area one hundred and eleven miles wide.

Accordingly, as this particular stratosphere journey to Chicago extends in north to northwesterly direction, there would be photographed nine luminous, globular, and isolated "bodies" on the direct course. And photographs made at an angle to the direct course would show numerous other luminous rounded and isolated "bodies," their number depending on stratosphere altitude and camera lens power plus the photographing angle. The intensity of gaseous sky content prevailing at the time of photographing would likewise influence the number of "bodies" to be detected by the camera lens.

The group arrangement of **Figure 4** is intended to convey how every luminous terrestrial sky area would appear; but such necessary illustrative grouping of sky areas does not permit the luminous sky areas to be separated, or isolated, as they will appear from distant observation. It should be understood that, when observed individually, the luminous curving-down of each depicted sky area causes it deceptively to appear separated and isolated as a distinct unit, or "body." No lens can detect and record more than one of the luminous disk areas at a given time. That feature, as previously shown, was proved by the U.S. Navy's rocket-camera photographs of luminous terrestrial sky areas over White Sands, New Mexico, and adjacent territory.

As the illustration's thousand-mile photographing experiment is in progress from New York City to Chicago, other similar experiments over the sky of corresponding thousand-mile areas can be moving in the stratosphere from Los Angeles and from Montreal, London, Berlin, Moscow, and

ch. ends p. 122

Rome. They would all be procuring *identical photographs* over their respective luminous sky areas. There could be variation in the quality and the quantity of light shading and distortion in some photographs over different sky areas. If the cameras of the different photographing expeditions possessed varying lens power, that would result in there being more or less luminous and isolated terrestrial sky-area "globes" photographed over different routes. However, if the same lens power is utilized in all cameras over all routes and if the same altitude is maintained, the photographic results will be approximately the same.

The qualification, approximately, is in order because conditions prevailing at the time of photographing some thousand-mile areas would vary with conditions prevailing elsewhere and with those of the thousand-mile area from which the numerical standard was developed. Gaseous condition of the various luminous sky areas could influence detection or mitigate against the possibility of detecting certain sky areas. The photographing angle would also contribute to numerical finding.

Thus, at this point one may have acquired some vague concept of the *deceptive isolated* terrestrial Universe that our luminous outer sky areas present to all observers from beyond the Earth. One needs but briefly consider the number of luminous isolated "globes" to be detected over a single thousand-mile area of the Earth's entire luminous outer sky surface. Naturally, the number of isolated "globe" to be detected can be expected to vary depending upon lens power, restricting angles of lens focus, and conditions existing at various terrestrial sky areas. In the latter consideration, stratosphere elements and gaseous sky content and expression would be factors.

It is reasonable to assume that a lens with greater power will embrace a wider terrestrial sky area than a weaker lens can. But the more powerful lens cannot detect as many "isolated globes" over a restricted sky area because of the fact that, by embracing a larger sky area, there will be an overlapping of the more numerous areas to be detected by the weaker lens. Where the weaker lens might show twenty or more isolated sky areas in one hundred miles of sky surface, the stronger lens might be expected to detect only ten or twelve, or even fewer.

However, the numbers here used are meaningless other than for comparison. No numerical accuracy is intended or required. The primary and broader purpose of the 1930 illustration was to express that all astronomical observations of so-called stellar areas are products of the inescapable lens deceptions which *must be* duplicated in every detail in telescopic observation and photography of luminous outer sky surface areas of the Earth. Realization of lens deceptions in the sky over our own backyard eloquently proves that telescopic observations of the celestial deal only with unrounded and connected celestial sky surface areas. And it is the individual concept which mistakenly *bestows the status of "globe" on celestial sky surface areas after the detecting lens has provided the area with a disk appearance.*

There should be great need for stressing this factor after three hundred years of mathematical astronomy which, in detecting some and conjecturing other luminous surface areas of the celestial sky, has developed the dictum extraordinary that the disk area of lens production is actually the "globe" which concept harbors. To avoid possibility of misunderstanding this paramount feature dealing with illusion and delusion, it may be further clarified as follows: The *unreal "globe"* which was sired by the *unreal disk* (because the lens alone was responsible for the disk) is astronomically established as a factual entity in the world of things. Is it not astounding?

Fortunately, current rocket-camera photographs of luminous outer terrestrial sky surface areas make it possible for

ch. ends p. 122

the first time in history to check and compare astronomical observations. That checking and comparing was denied to telescopic observation for many centuries. And it has since been denied to astronomy's hired assistants, telescopic photography and spectroscopic analysis. However, it has now proved the complete fantasy of isolated globes or spheres "circling or ellipsing in space."

Though the unprecedented opportunity for checking and comparing assumed conditions of celestial finding with factual conditions of terrestrial finding is now available to astronomy, it is questionable if the astronomical fraternity will take advantage of it. "We see only that which we want to see. And we believe no more than that which we want to believe." Hence, primed observations are as dubious as spies. Nevertheless, though primed observations may be known to be so untrustworthy, such primed observations are retained as companions because that seems to be the easiest course. To reject them would impose an effort and a responsibility.

Since rocket-camera photographs have established that the deceptions of lens function are inescapable, it follows that, once the telescopic finding is accepted at its face value, deducing robots instead of human astronomers may as well check the lens findings. What the astronomers may interpret of the telescopic-photographic plates becomes entirely irrelevant, if the lens error reproduced on the plates is accepted as fact. Alas, the astronomer seems to be painfully reluctant to admit that proof of the error is at hand.

It is pertinent to explain that the identical spectrum variations of celestial analysis will be found to apply to luminous outer surface sky areas of the Earth. The same misinterpretation of values will ensue. And with realization of the terrestrial sky areas' factual values, the misinterpretation of celestial values should become manifest.

Though terrestrial sky areas are known to be continuous and holding their allotted place in the Universe structure, their billowing or fluctuating within the cosmic area of their original construction and placement will be accredited the same fantastic motions astronomically prescribed for the so-called "stars" and "planets" of celestial sky areas. When terrestrial sky areas are analyzed from the same distance and with the same astronomical equipment, their gaseous content and movement will produce all that which celestial sky gas produces for spectrum analysis of terrestrial astronomers. However, from our celestial observatory we would not dream of interpreting the spectrum recordings as astronomers now interpret the recordings from celestial sky areas. With knowledge of our terrestrial sky we would know better.

Thus, returning to the illustrative thousand-mile course of terrestrial sky-light illusions, we find that the stratosphere journey from New York City to Chicago at an altitude of one hundred miles or more must develop the following observational and photographic conclusion:

The deceptively globular and isolated luminous sky areas would require seeing the "planet" of New York City. Then, in the order designated, there would be seen the "star" of Albany and the "planets" or "stars" of Utica, Syracuse, Rochester, and Buffalo. Then at an angle from the main line of perpendicular observation over the cities of New York State would be observed the "star" of Erie, Pennsylvania. As the course continued toward Chicago there the "planets" of Cleveland and Detroit would loom. Other vague "star" scatterings would be observable in all directions away from the direct course being photographed on the perpendicular.

Every thousand-mile area of the luminous terrestrial sky would present the same deceptive appearance. And the sky areas would show corresponding celestial sky variations of luminosity due to variations *of the chemical content and gaseous activity* of the respective terrestrial sky areas. (Though this may be repetitious, it should here be explained that the familiar blue sky's varying depth, or blueness, observable

ch. ends p. 122

from time to time and from place to place at the same time, actuates variation of the outer sky's luminosity.)

The following feature also serves as an agent for the lens-developed illusions of record. The torrid equatorial and the frigid Arctic and Antarctic sky areas would be shown to possess marked difference in the depth of their luminosity when compared with the luminosity of Temperate Zone sky areas. That would mean very little if the Universe whole contained but one torrid and two frigid zones as now known at terrestrial level. However, the zones of the terrestrial are duplicated over and over again throughout the Universe whole. That factor influences difference in light waves and colors now registered from luminous sky areas of the celestial which are otherwise of the same composition. Corresponding differences for corresponding reasons would be shown to develop from terrestrial sky areas.

Were we to increase the hundred-mile altitude to five thousand miles, the sky area of the illustration's course from New York City to Chicago would loom as a wide layer of "stars." Then, as our telescope was adjusted at an angle for observation of the sky territory northeast of New York City, there would be detected sky area "stars" of Connecticut, Rhode Island, and Massachusetts. The number of "stars," "star clusters," and "double stars" to be detected over that sky area would depend on lens power and other conditions previously described.

The extent of our stratosphere search for terrestrial sky "stars" could continue over the Atlantic Ocean beyond Boston. "Stars" detected at such points would represent the rim of the terrestrial "stars" area first detected at New York City. And detection of "stars" would not be restricted to a direct eastern area. As it embraced the area from New York City to Boston in an easterly direction, it would also embrace a wide area in a northerly direction to the Canadian border and south to the Gulf of Mexico.

Under telescopic observation some sky areas would become vaguer, while others of the same area would be more luminous. The more luminous might appear at the Atlantic Ocean rim, and the vaguest might be detected in nearby Connecticut. Other sky-light areas would appear so vague as to make for the determination that no sky luminosity, and therefore no sky, existed at such points. The detection of sky luminosity, celestial and terrestrial, does not depend solely upon distance from the observing point.

North, east, south, and west, our terrestrial sky-light would reproduce that which is presented by celestial light. The extent of our view of the Earth sky's "Heavens above" would depend on the angle of observation in the stratosphere, the power of the detecting lens, and the gaseous condition of the most remote sky areas at the time of observation. At altitudes of one thousand to five thousand miles in the stratosphere, the most powerful telescopic lenses and their companion camera lenses would likewise record all the grotesque entities presently recorded of the luminous outer sky surface areas over the land of other parts of the Universe. Such recording would be of sky areas over the known land of New York, Connecticut, Rhode Island, and Massachusetts, as well as over the water of the Atlantic Ocean. Hence they could readily be determined as the illusory condition considered to be real when the same entities are observed over celestial land areas.

The astronomically recorded "Horse's Head in the great nebula in Orion"[2] and "the spiral nebulae in Cygnus"[3] would be reproduced in certain terrestrial sky areas where the play of sky gases plus lens magnification would develop such *gaseous formations*. And if the light distortion appearing

[2] The Horsehead Nebula, also known as Barnard 33, is a dark nebula within the Orion Molecular Cloud Complex, appearing as a dark cloud shaped like a horse's head.

[3] The constellation Cygnus is home to the Cygnus Loop, a large supernova remnant also known as the Veil Nebula.

ch. ends p. 122

as a dark form in the terrestrial sky area was not defined as the "Horse's Head" in the celestial sky-light of astronomy's Orion, it could readily be designated something else related to horse anatomy. Such designation would not obscure the fact that it is nothing but sky-light distortion.

That which applies to the dark formation in luminous sky area likewise applies to the white formation in the astronomical "nebula of Cygnus." The ectoplasm-like white veil, or film, of the Cygnus sky-light area will be duplicated in terrestrial sky-light. It may be found to develop in the sky-light making the "stars" of Portland, Old Orchard, and Kennebunk, Maine. Or it could as readily be observed in the terrestrial sky-light "star" of Kalamazoo, Michigan. That sky-gas condition which astronomy is pleased to describe as the "nebula of Cygnus" *has already been photographed in the luminous terrestrial* sky over White Sands, New Mexico. And it could be reproduced in ever so many terrestrial sky areas under conditions favorable to its formation.

Another interesting observation from the haloed realm of astronomical deduction is that dealing with the "nebula M-31 in Andromeda."[4] Though it is conveniently mathematized as being thirty-five hundred million times the *weight* of the Sun, it can be readily *dissipated* under lens magnification. This expresses the ultimate of abstraction in the application of abstract mathematics. Despite such estimated sky-gas weight, a telescopic lens can dissipate the so-called "nebula" formation. Yet the lens cannot penetrate through the gas density to the underlying land.

The depth of abstraction becomes evident as one realizes that there cannot possibly be an authentic gauge for the Sun's mass. And any weight estimate is absurd. Though it fits the Universe of illusion, it can have no application to the Universe of reality. It is comparable to an estimate

---

[4] M31, also known as the Andromeda Galaxy, is a spiral galaxy located in the constellation Andromeda.

concerning the birth and ancestry of God. One need not burden conceptional capacity in a forlorn attempt to determine the meaning of that figure thirty-five hundred million times the weight of the Sun.

Regardless of how one cares to view the application of abstract mathematics and the real meaning of so-called "nebula," the paramount fact remains that no sky-gas motion, seeming or real, has any bearing whatever on the realistic connected land existing under all sky areas. For reasons abundantly disclosed, sky areas must be considered isolated. The art of astronomy, though impotent to penetrate the gaseous celestial sky envelope, regardless of what its density may be, is restricted to observation and analysis of luminous sky areas and the movement of their gases. And astronomy's failure to grant that "nebula" is an aspect of sky-gas motion fosters gross misinterpretation of cosmic values.

Experimentation proves that in observation of light and luminous areas there will at times be formed grotesque creations. At other times the formations will be dissipated. It depends considerably on the angle of observation, the gaseous movement of the luminous area at the time of observation, and the amount of magnification of the light or luminous area.

Microscopic observations clearly express such features, though there exists in microscopy a possibility of error which is infinitesimal in comparison with the unlimited possibilities for telescopy. Observations of a microscopic field establish that too much magnification of the field's specimen will cause it to be obscured, whereas a different light quantity will distort the specimen.

Hence in a factual study of lens capriciousness[5] it is established that the important feature is not so much what

[5] Lack of consistency.

ch. ends p. 122

is observed but, rather, *how and under what conditions observations are made.*

In spite of claims to the contrary, abstract mathematics and their competent mechanical aids and guides can in no way correct the structurally inherent lens culpability.[6] Size and power of a lens has nothing to do with the error of lens principle. A thousand-inch lens cannot eliminate the error, but it can and will *magnify the error.*

From the enviable thousand-mile observation point in the stratosphere, the "Heavens above" would be observed everywhere and at every angle of observation. Every luminous outer sky area over the entire Earth, or as much of the Earth's sky that could be detected, would present a vista of the "Heavens above." The terrestrial appearance in no way differs from that of celestial sky areas observed from the terrestrial. The rhythmic shift of light motion within some luminous outer sky areas of the terrestrial would also present the same characteristics under spectrum analysis as presently found in the light of celestial sky areas. And that corresponding activity would cause it to appear that the "star" of East St. Louis, or some other terrestrial sky area, would be *burning up its terrestrial orbit* at a devastating rate. And it would *deceptively appear to be circling toward our observation point* in the stratosphere.

On the other hand, it might appear to be as rapidly receding from our position and away from its normal location. The appearance of approaching or receding would depend upon the intensity and motion of sky gases at that particular place when observation was made. Such condition would deceptively appear when in fact nothing was going anywhere, either toward or away from our stratosphere observation point.

Some terrestrial sky areas would seem to flicker, or

---

[6] Responsibility of fault or error.

fluctuate. The motion of some areas would appear to be constant and therefore imperceptible as motion. The motion of others would appear to be variable. And the constancy or variability of terrestrial sky-light motion would correspond to that recorded by the light curves from celestial sky-light areas. However, with the physical knowledge possessed of our Earth's sky, no reasoning person could ever ascribe to such motions of terrestrial sky-light that which astronomy interprets from identical motions in celestial sky-light.

Celestial and terrestrial sky luminosity and the motions of such light have a common heritage. They are of the same Universe family. Further, one is as continuous with the other as the circulating blood of the human body which actuates the left side as well as the right side, and thereby nourishes the entire body.

Mathematical astronomy has not, and will not, detect that obvious continuity feature from lenses and figures. That feature, being of the Universe of reality, was not entrusted to the uncertainty of abstract figures and symbols. Though such figures and symbols are endowed with precision and positiveness, the endowment applies to and benefits only the unreal mathematical Universe.

In analysis of light waves from various so-called "star" areas of the Universe at times two spectra[7] are observed to move back and forth. They prescribe, or there is prescribed, a waving or undulating motion of the sky-light under analysis. The astronomer's conclusion must be that such duality of motion presupposes dual entities in motion. He does not consider the motions attributable to sky gases. If he did, he would be empowered to consider many other features this book contains. Instead, when spectroscopic examination confirms the dual motion, the astronomer must assume that confirmation has been had of two distinct entities, or

[7] Plural of spectrum.

ch. ends p. 122

"bodies," whereas in reality all that the astronomer's eyes, the telescope and camera, the spectrum and the spectroscope, have established is that *dual motion* is taking place in the celestial sky-light area.

It should be further noted that none of the observations and tests have anything to do with land areas of the Universe underlying the sky-light being tested. They are restricted to a determination of celestial sky-light content and activity. They are impotent to deal with the land existent under the sky-light. Though there is land under all celestial and terrestrial sky-light, there is no "body" in motion, to say nothing of two separate "bodies" in motion. The ever-active sky gases are responsible for all detected motion. Other factors may influence the portrait of motion which the lens detects. They also influence the spectrum.

Hence it is nothing more formidable than the misinterpretation of sky-gas motion which leads to the conclusion of "spectroscopic binaries," or "double stars," in this particular instance of celestial sky-light analysis. Duality of gas motion can exist. But duality of "bodies" can never exist, for the reason that there are no celestial "bodies" to have motion.

That particular astronomical feature was embraced by the original treatise *Physical Continuum* as early as 1927. There it was disclosed that every sky area of the Earth deceptively appears to be circling or revolving. That 1927 claim has application to the entire Universe. It discounted astronomical interstellar space and the circling or ellipsing of assumed isolated "bodies" in restricted space orbits. Orbits are definitely not required for the motions of luminous sky gases over land areas that are connected throughout the Universe and are not "circling or ellipsing in space."

Energy in motion is restricted to waves of varying length and intensity. All of modern enterprise establishes that feature of natural law. And the active sky gases of the terrestrial and the celestial conform to the principle of motion. What deceptively appears to be happening should be known as illusory by modern astronomers. Then would they be able to discount the seeming celestial conditions which perpetuate and enlarge upon the primary illusion developed by lens function.

Another of the many extraordinary features of astro-mathematical confusion is that which grants so-called "nebula" centers composed of gas, and then proceeds to mathematize that such gas is formative in the stratosphere as the nucleus of "star" matter. Such a wayward conclusion results from the fact that the central regions of some luminous celestial sky areas under observation defy penetration and dissipation of their light by the most powerful lenses. Hence such concentrated central points are mathematized and assumed to be something different and remote from the remainder of the sky area.

This observation is one that brings mathematician astronomers to the door of reason. But, alas, they refuse to enter. Such observation should show that the substance, deceptively appearing to be formative in the stratosphere and apparently alienated from the central luminous sky area, is gas movement of the sky-light area. In a case of this kind the astronomer comes so close to the truth that it is painful to realize how his misconception of values demands that he adhere to the faulty premise and forsake the truth so glaringly presented.

It seems that something pertaining here was mentioned about two thousand years ago by the immortal Master of parables, who pronounced: "None are so blind as they who will not see."[8]

That intensification of sky-light in some areas, celestial

---

[8] The first known instance of this phrase is attributed to English playwright and poet John Heywood (1497–1580), which can be found in his 1546 book of proverbs as: "Who is so deafe, or so blynde, as is hee, that wilfully will nother heare nor see." However, the phrase has its Biblical roots in Jeremiah 5:21, Ezekiel 12:2, Matthew 13:13, and Mark 8:18.

ch. ends p. 122

and terrestrial, is a very natural condition. And it is related to the following. As one looks at the massed luminous coals of a furnace fire, the fire's luminous area, with the exception of the center, may under intent observation be broken up into viewable formations. The center, in holding the concentrative force of the fire and emitting the greatest light, must defeat any effort to see it as other than a vast concentration of impenetrable light and heat. Nor can its light be dissipated. If the observer of such a furnace fire were at sufficient distance, and if he had not had direct experience with such accumulation of heat and light, he would be compelled to conclude that the border areas of the fire concentration were *different in substance and detached from the central area.* Yet composition of the central area would be no different than the fiery matter viewable at the extremities of such an intense luminous center. Every area would be continuous with the center.

Nevertheless, astro-mathematical calculations develop the fallacy that the center luminosity, not amenable to lens dissipation, is of a different model and is isolated from the extremities of that same center. Actually, the central concentration of a luminous sky-gas area bears the same relation to the remainder of the detected sky luminosity as the furnace-fire center is related to the extremities of the fire accumulation.

The problem is resolved as follows. The gas mass of average sky-light is readily lens-detected in "star" proportion, whereas extraordinary gas mass content prohibits lens dissipation of the sky-light. Accordingly, there can be lens detection and "star" formation only of those parts of the luminous area having less concentration than the central area. Hence the center invites the conclusion that it is a remote "body." The furnace-fire center invited the same conclusion. Hence the entire area is a "nebula." And in a

Universe of reality, any "nebula" is but *an aspect of luminous sky gas and light* manifested throughout the entire Universe.

The fascinating feature of sky-light formations from sky-gas motion becomes a double feature as we review the 1946 rocket camera's accomplishments. On that occasion a corresponding "cloud" formation was photographed *within* the luminous sky area over White Sands, New Mexico. It was also erroneously claimed to be a formation in the stratosphere. Hence if it were of stratosphere formation, the astronomer would have to consider it a "nebula."

Now it may be seen that something is about to happen to the mathematical astronomer's abstract figures and symbols of distance. It is to be recalled that such a white cloud-like formation was developed at the feeble distance of ninety miles from the stratosphere photographing point. It should also be recalled that a "nebula" is the assumed mathematico-astronomical substance supposed to be observed only at vast distances and because of distance. It is supposed to be the stuff of which "stars" are made. Hence "stars" are being made in our own backyard. Marvelous Creation!

Observe again how dangerously close astronomers come to the answer contained in lights of telescopic detection. Yet they will not see that the so-called "nebula" is part of celestial and terrestrial sky-light and that its detection anywhere is an expression of sky-gas function.

It matters little if the original claim concerning the stratosphere photograph is retracted. It holds that a white area of the photograph is a "nebula" in the stratosphere and that the white patch was detached from the remainder of the photograph of terrestrial sky. The self-evident fact is that such a formation cannot possibly be considered remote from the remainder of a sky area photographed at a distance of only ninety miles. If it were remote it would not have appeared as part of the photograph, as it did.

Whatever determination is made of that white patch in

ch. ends p. 122

the luminous sky area over White Sands, it demands the discard of at least 50 percent of astro-mathematical deductions concerning the structure of the Universe and what is taking place throughout the Universe. What that U.S. Naval Research Bureau rocket camera developed cannot simultaneously be considered "nebula" and "not nebula." It cannot represent something possible only at unfathomable distance and at the same time be proved to exist in terrestrial sky-light less than one hundred miles away from the photographing point.

While such sky-gas formations were concluded to exist only at assumed distance the mind cannot grasp, and while they were assumed to be celestial entities unrelated to terrestrial sky-gas development, the astronomical conclusion that they were detached from luminous celestial areas had to be accepted. Hence they were undisputedly established as elements of the astronomer's so-called "interstellar space." And with the assumption that they were contained in that space rather than in the detected celestial light, they were assumed to be building material for so-called "stars."

The singular feature of this immediate exposition is that the astronomer, by concluding that his so-called "nebula" is building material for "stars," moves in a centuries-long course toward admission that Physical Continuity is a reality. But the astronomer does not know he has admitted it. If the astronomer's "nebula" builds "stars," it is gas accumulation. And Physical Continuity shows how terrestrial and celestial sky-gas accumulation must ever be considered "stars" and "planets." However, the meeting of theory's abstract course with the course of reality here described would prove to be too simple for complicated astronomy.

It should here be related that when the early Universe interpreters prepared the foundation for the elaborate astronomical framework they could in no way anticipate the rocket's development and its sensational performance.

Its camera's stratosphere photographs have shattered considerable of the suppositive[9] astronomical fabric. And as rocket camera photographs have been responsible for such magnificent accomplishment, they have brought the realistic celestial structure much closer to the terrestrial. They have also accentuated the pace of modern man's conquest of the universe about us.

In the foregoing reference to lens penetration and dissipation of celestial sky-light concentration, the word "penetration" implies only lens ability to grasp such luminous area for the purpose of recording it. It is a case of penetrating into the luminous surface *but not through* the light accumulation of any sky-light area, celestial or terrestrial. No lens can be expected to penetrate through sky-light at the abstruse[10] distances conjured by astro-mathematics.

Particularly does such apply when it has been conclusively proved, by the stratosphere ascensions of 1931 and 1935 and by rocket-camera photographs since 1946, that sky-light *cannot be penetrated through* at a distance of less than one mile in the first case and at ninety to one hundred and ninety miles in the latter case. Hence the necessarily oft-repeated description of lens function must hold. No telescopic lens can *penetrate through* celestial or terrestrial sky-light and detect the underlying land. If lenses could so perform, and if their findings could thereafter penetrate certain interpreting substance, all the celestial problems would have been resolved when the first telescope was fashioned.

This account of lens failure to penetrate through light presupposes an absence of the light-penetrating emulsion applicable to photography. If there is a medium applicable to telescopy, it represents a very recent development and is unknown to this writer. However, even with application of such a light-penetrating medium to the camera lens, the

---

[9] Presumed to be true or genuine, especially on inconclusive grounds.
[10] Obscure; difficult to understand.

ch. ends p. 122

area photographed through light must be distorted, and use of the medium will be readily detected by evident distortion of and foliage on the Earth's surface.

The greatest boon to mankind, other than the secret of overcoming death, would be the invention which might permit telescopic observation of that which is under every light detected in the universe about us. Then this volume might not be necessary.

In what might be considered a capitulation to reason, there is observed the measure of penetrating into, *but not through*, the celestial sky-light surface astronomically designated "nebula M-31 in Andromeda." That celestial sky-light has already received some attention here. Though the land exists under such a sky-light area, there can be no land consideration by astronomy, which deals only with the outer surface sky-light. A much different story would be unfolded by astronomy if telescopic lenses could penetrate celestial sky-light, particularly at the distances supposed to be involved.

At the aforementioned celestial sky-light point, the mathematical astronomers estimate a "nebula" accumulation weighing thirty-five hundred million times the Sun's mathematized weight. If one dotes on figures, such figures should be impressive even if no light or Sun existed for the figures of comparison. As such colossal figures are presented, it is asserted that the "nebula" mass can be dissipated under lens magnification. However, in this instance, the manner of dissipation merits qualification. No area of sky gas is dispersed by a lens, but the fact of seeming dissipation is sufficient to establish that the telescopic lens detects nothing but luminous sky gas.

This dissipation in no way implies penetration. It is but a superficial dissipation likened to the dissipation of an impenetrable fog bank experienced on the Earth's surface. Though the fog bank is not to be penetrated by the optic lens, its outer areas may in various manner be dissipated. If the fog bank could be dissipated at our will, it would not be impenetrable. If it could be dissipated in the true sense of the word, we could see beyond it. Therefore, we could not say the fog had been penetrated.

The telescopic lens cannot and does not dissipate sky gas to permit penetration. Were such possible, the land underlying the sky gas would be detected. But since astronomical conclusions do not seem to approximate such reasoning, we will pursue astronomical deduction as the astronomer would have it:

1. This assumed "nebula" mass, which is really sky-gas cloud over a celestial land area, is mathematized as being thirty-five hundred million times the unknown Sun's mass and weight, assumed to be known through the same abstract mathematical procedure.

2. And the light of such a "nebula" mass can be dissipated, but not penetrated, over a cosmic distance assumed to be only a feeble nine hundred thousand light-years. This impressive astronomical light-year is the distance a ray of light is assumed to travel during the course of our known year of three hundred and sixty-five days while moving at the speed of 186,000 miles every second of that year. That yearly distance is a trivial *six trillion miles*. Now that single light-year distance need only be multiplied by nine hundred thousand.

Though one cannot possibly conceive a fraction of such distance one may now easily realize precisely how a telescope lens can detect and dissipate light existent at such distance. One may also have full realization why the lens cannot penetrate celestial sky-light.

Caution seems to dictate that one not attempts to visualize such distance or the manner whereby a telescope lens

ch. ends p. 122

might detect and dissipate light over such an inconceivable distance, yet lack the power to penetrate it. Though there could be double, triple, or a trillion times such inconceivable distance to infinity, there is no lens created and none that could be created to detect light over a distance mathematized as a small fraction of one light-year, to say nothing of nine hundred thousand light-years.

Such distances do not exist for realistic entities in a world of reality. They exist only in and for the abstract Universe of the abstract mathematician.

A ray of light is most factual. A telescope lens is a realistic entity in spite of its inherent error. And the established function of light ray and telescope lens prohibits the fantastic performance as mathematically prescribed. The prohibition is proved by the fact that a lens is compelled by its function to create curves in its detection of light. And rays of light are compelled by their function to wave and bend as the curve-producing lens seeks to detect them. The lens does not penetrate through six trillion miles of space before developing the curve, and the ray of light does not travel such distance without bending.

The lone factor of lens curvature prohibits such telescope accomplishment. And the abstract determinations have been dictated through the control of abstract mathematics. They are sole arbiter of the situation, qualitative and quantitative.

Were one competent to imagine a telescope lens of such construction as to eliminate lens curvature, and thereby to permit lens penetration of boundless inconceivable infinity, by what reasoning could it be known that inconceivable infinity had been penetrated to its limitless extent? Were we to grant conceptional ability to retain other than by mathematical symbol a time space end to infinity, what would we name that which would extend beyond the finite bounds of infinity? Regardless of designation, would it not constitute a continuance of infinity?

The human mind in wayward fashion seeks to establish the end, though it must ever be denied knowledge of the beginning. The empty procedure is likened to a forlorn attempt to determine the Creator's creator. Therein it would be found that, when mind established a power behind and preceding the Creator, the mental process to establish First Cause to supersede the mind's designation of Creator's creator would develop into an endless and futile procedure. And mind in its quest would become lost.

The ultimate of abstract astro-mathematical endeavor defeats the purpose of all educational advancement and modern scientific research. The endeavor reflects the immature wisdom of the child in Sunday school class who being told that God created the world, was impelled to ask, "Who made God?" Astro-mathematics rush headlong toward the elusive end of the Universe mathematically ordered. In so doing they deny the Universe of reality at hand. And in that denial they demand that modern man relinquish his divine right to conquer and to inhabit the resplendent universe about us.

Like the child who should first seek to know God and His abundant manifestations close at hand, the astro-mathematician should first seek to know the meaning of cosmic manifestations before attempting to find the end of the Universe. Somehow there seems to be more glamour attaching to the second course — and, like most glamour, it is shallow and unproductive. No portion of the astronomical portrait dealing with the so-called "nebula M-31 in Andromeda" has application to a Universe of reality. As the astronomer presents it, the portrait is one which applies in its entirety to the unreal Universe of abstract mathematics.

The lack of realism in astro-mathematical conclusions may be understood from the following. If from the nearest celestial point from San Francisco, London, Rome, or any other terrestrial point there was erected an astronomical

ch. ends p. 122

observatory equipped with the identical mechanical equipment and astronomer deductions now applying to observations of the celestial, the conclusions to be reached in observations of the terrestrial would compare with present conclusions concerning the celestial. The distances estimated from that celestial observatory to luminous terrestrial areas would have to allow for the space assumed to exist between apparently isolated areas of the terrestrial. The fictional space pattern now applicable to and influencing distance estimates for celestial areas would have identical application to the assumed "interstellar space" between apparently isolated terrestrial "bodies."

Never could the Earth territory of the Universe be seen as a single unit in space, but only as popular misconception has held. Lens curvature prohibits any such distant observation. And lens curvature demands that the Earth be seen as the multiple globular and isolated "bodies" deceptively arranged for the celestial. The absurdity of the astronomical estimate of the sky gas mass in that area the astronomer knows as "nebula M-31 in Andromeda" would be established by corresponding appearances in areas of the terrestrial sky. The apparent gas mass of at least one area of the Earth's entire luminous outer sky surface would be found to present the same appearance as the area known as "nebula M-31 in Andromeda," and if its assumptive weight were to be compared with the Sun's assumed mass, the figures applied to the Andromeda condition would hold equivalent application in the world of figures.

Moreover, the inconceivable distances involved in the detection of the Andromeda sky-light could be made to apply to known areas of the terrestrial sky only a few thousand miles away from the observation point. The factors heretofore described, particularly the assumptive space factor, would permit of the most abstruse mathematics in description of distance.

Were we to establish at a ten-thousand-mile stratosphere altitude an imaginary terrestrial sky line as the measuring base through our terrestrial sky-light areas, it would be considered to represent the terrestrial "star" area conforming to Herschel's base formula for celestial sky-light areas.[11] There would thereby be formed a terrestrial "galactic system" agreeing with the present celestial "galactic system" of astronomical order. It would embrace terrestrial sky-light areas to a mathematically designated extent in all directions away from the terrestrial "galactic plane."

Now, it must be understood that the distances presently recorded from the celestial "galactic plane" to the greatest extent of celestial sky-light detection are purely attributes of mathematical formula. They are most unreal.

Then, in applying the customary astronomical yardstick, the presently known and real distances from the terrestrial "galactic plane" to the most remote terrestrial sky-light points would demand the identical abstruse distance consideration applicable to celestial sky-light points detected beyond a given distance from the celestial "galactic plane," or line.

Sky-light points of a known twelve-thousand-mile terrestrial sky area, representing one half of the determined Earth's circumference, would have to be considered millions of miles away from the terrestrial dividing line and from the observing point only ten thousand miles away. Were observation made from the celestial Moon point three hundred thousand miles away from the terrestrial, the most remote terrestrial sky-light points from the terrestrial "galactic plane" would have to be any number of light-years away from the observation point. That purely mathematical consideration for a mathematical Universe would apply even

[11] German-British astronomer and composer Frederick William Herschel (1738–1822) used a method called "star gauging" to understand the structure of the Milky Way galaxy based on the distribution of stars and their apparent brightness.

ch. ends p. 122

though the most remote terrestrial sky-light points were actually embraced by the Earth's known circumference of twenty-four thousand miles.

These absurd conclusions in application to the terrestrial conform to astronomical conclusions concerning the celestial. And the greatest contributor to that absurdity is the assumed space between all terrestrial sky-light points detected from the terrestrial "galactic plane" to the most distant terrestrial horizons. Though we know the terrestrial sky is as continuous and spaceless as the underlying terrestrial land, the illusory space would be an important factor causing enlargement of distance to an incalculable extent.

In conjunction with the terrestrial-sky space illusion, terrestrial sky-light gas expansion and contraction and sky-light radiation and the additional illusion it imposes would likewise contribute to an unreal distance pattern corresponding to that astronomically ordered for the celestial. The speed of light through the more realistic medium of infinity's perpetual darkness, as opposed to the speed of light assumed from man's artful but artificial experiments at sea level, is another factor. These and numerous other purely technical but extremely important elements are the influencing agents in compilation of astronomical data having no application whatever to the celestial of reality. Their influence extends to the terrestrial and the terrestrial sky's natural manifestations. They, too, must be misinterpreted through abandonment to the illusory.

The celestial and terrestrial comparisons, now proved to be of merit as a result of stratosphere ascensions and rocket flights, are here afforded timely expression. They show the terrestrial sky-light formations and deceptions already encountered in stratosphere photographs of luminous terrestrial sky area. Such photographs attest to Physical Continuity no less than the land extent continuing beyond theory's North Pole and South Pole "ends" of the Earth. One feature complements the other. And they jointly contribute to the development of a new and factual portrait of the universe about us.

The little publicized radar portrait of a substantial area of the celestial sky also contributes to the Universe portrait. And such features, collectively, establish beyond any doubt that the realistic pattern of the Universe is diametrically opposed to that developed by astro-mathematical deductions of the centuries.

If one finds it difficult to accept these Physical Continuity dictums in spite of physical proofs sustaining them, the following should be considered. In a child's mind may be fixed the deluding features of the "Fable of the Stork."[12] The child lacking knowledge of procreation, must cling to that fascinating fable. The fable must prevail if the child's mind is not sufficiently developed to comprehend the meaning of reproduction, with its successive stages of cell transmission, fetus development etc. The child's mind may even acquire the accepted descriptive of birth. The child may be able to express the words *sex, born, baby, growth,* etc. It may even witness the moment of a birth. Yet as long as the immature mind is dominated by the image of a long-legged bird delivering babies, it may behold a million babies and remain in ignorance of how they arrived.

That child's mind differs not from the undeveloped adult mind. Though the adult mind certainly knows how babies are delivered, it can remain as closed as the child's mind concerning other features of life and of the Universe. That which concept does not hold is beyond the bounds of possibility for both child and adult.

As it is with the child's mind, so it is with the astronomer's

---

[12] Popularized by the story "The Storks" by Danish author Hans Christian Andersen (1805–1875), the fable is a whimsical tale where storks are depicted delivering babies to families. However, the story is rooted in Greek mythology, where a vengeful goddess named Hera transformed a woman named Gerana into a stork, who was often depicted with a baby in her beak.

ch. ends p. 122

mind, which causes him to express the words *curving, waving, bending, fluctuating,* and *undulating.* They should afford ample knowledge that creative energy at work does not circle. And they should be a key for understanding that globes or spheres do not comprise the celestial or the terrestrial. Yet, despite the astronomer's broader observation and deeper calculations of luminous celestial sky gases in motion, he demands that unseen mass "bodies" be prescribing all motion, and the wrong motion.

The undeveloped child could be shown realistic pictures of baby delivery and, through domination of the fable, remain ignorant of reality. So it is with the astronomer who, in viewing physical proof at hand of the fallacy of "isolated bodies," persists in clinging to the "star" and "planet" fable. And he makes every effort to fit proofs culled from a world of reality into his world of illusions. The illusory must be preserved at any cost. It is the astronomer's truth.

There is not a feature of telescopic observation and photography, and of spectrum analysis, considered applicable to the universe about us which does not apply with equal force and volume to corresponding tests of the Earth's outer luminous sky surface. Yet . . . modern enterprise has established that such absurd features are purely illusory. And they do not apply.

All the fantastic entities assumed to exist throughout luminous celestial sky areas seem to exist in like observation and analysis of the constantly shifting gases of the Earth's sky. And it must never be forgotten that all observations, analysis, and resulting conclusions apply only to the *sky-gas energy* of celestial and terrestrial sky-light areas. There is no application whatever to the land under such sky-light areas.

The cosmic agency which contributes to the many deceptive movements of the least luminous and the most luminous sky areas is responsible for the light shifts, fluctuations, and undulations. And it thereby indirectly governs the resulting grotesque formations so deceiving to the observer. The cosmic agency and creative force, beyond astronomy's embrace, is cosmic-ray activity. It is constantly bombarding every outer sky area of the entire Universe. The rays have no directional pattern. They are not restricted to any course or channel in their ceaseless movement throughout the infinite realm of darkness, of which our immediate stratosphere is a part.

Sown by the Master Planter, they are strewn from the Sun's impenetrable crater in a seeming helter-skelter. And in such apparent nonconformity to pattern, they establish the most profound creative pattern. Moving with immunity to man-made laws applied to the Universe, they affiliate with receptive outer sky areas everywhere along the celestial and the terrestrial course. They charge one sky area and supercharge another with their magnetic force. As their force is concentrated on a particular sky area of the celestial or the terrestrial, there is developed within that sky area an unprecedented accentuation of customary motion which befuddles distant observers. In other sky areas and at the same time, the dispensation of that creative solar energy remains stable in a perfecting balance of the whole Universe sky. But concentration of force upon one sky area exerts a measurable influence on neighboring sky areas.

Hence there is produced for the bewitching of mortal mind a unique series of motions within luminous sky areas under observation. But whether such motions are real or fancied, they are always motions of the sky. Never are they motions of the realistic land, which, though unseen, is always present under the sky-light.

Reason dictates that one does not erect a roof unless one is to have a house under the roof. The roof is the protecting medium for all the wood or concrete structure underlying. The roof is symbolic of the structure. And the magnificent but deceptive lights of astronomical observation and record

ch. ends next p.

are areas of a creative roof which cannot be seen as a collective and continuous whole for the reasons explained here. Our terrestrial sky covers our room of the Universe House in the same manner as every so-called "star" and "planet" covers the endless celestial rooms of the same house. Our sky, in common with all celestial sky, cannot be observed as a connected unit. It likewise presents to distant observers the identical pattern of varying luminosity and motion that we observe of the celestial sky. The astronomer expresses that factual sky-light variation of the celestial roof as "star magnitude." And that term is synonymous with "sky-light intensity."

That causative activity, of which so little has been learned, performs other wonders implied by the late Dr. Robert Andrews Millikan's memorable announcement: "Creative Life Force is at work throughout the entire Universe."[13] But the wonders of that Force at work are not to be determined by abstract figures and symbols of figures.

---

[13] Based on our research, rather than being a verbatim quote pulled from a source, this appears to be a general paraphrasing of Millikan's assertion that cosmic rays are products of continual creation in the universe.

7
"On Earth as it is in Heaven"

In **Figure 5**, the U.S. Naval Research Bureau's V-2 rocket camera photograph of a luminous, deceptively globular and isolated-appearing area of the Earth's outer sky from an altitude of one hundred miles over White Sands, a white cloud-like formation appears in the luminous sky area. It will be recalled that the formation, resulting from light variation within the luminous sky area photographed, was misinterpreted as a cloud in the stratosphere.[1]

Consider what the same white formation would be conjectured to be at a distance of twenty thousand or one hundred thousand miles. There can be no question about the astronomical label: it, like many corresponding celestial sky-gas formations, would have to be known as a "nebula" adrift in the enveloping stratosphere sea of darkness. That description would apply despite the fact that the white portion is in reality an intricate part of the luminous sky areas.

Black patches detected in the so-called "Milky Way" section of the celestial sky are intriguing partners of the white patches. They would also be detected in the dense center of our terrestrial sky where sky-light intensity presented to telescopic observation a "richness of star field."[2] That

[1] See Preface photo.
[2] The density or abundance of stars in a particular area of the night sky.

terrestrial sky center would depend on the observation position held in the stratosphere or on a celestial land area.

Were we to change our present terrestrial location to that celestial location now considered the "Milky Way," it would be found that the terrestrial sky over the land position we left holds the greatest concentration of sky-light points, and that terrestrial sky section would merit the designation "Milky Way." In comparison with other terrestrial sky areas, it would *seem* to hold more light points. But because there seemed to be more, they would individually appear to be much less luminous than other sky-light points detected singly. Or, if the sky over the particular terrestrial point of departure were to lack the apparent profusion of light qualifying it for celestial "Milky Way" comparison, other terrestrial sky areas would possess requisite seeming profusion of light points. Hence across the luminous stretch of our entire terrestrial sky there would be found from distant observation at least one sky-light area corresponding to the celestial "Milky Way."

As our angle of observation away from the *overhead* terrestrial "Milky Way" was accentuated, it would be found that there was a *seeming* diminishing of sky-light concentration or, as astronomically defined, a modification of the "richness of the star field." Though the astronomically defined "richness of the star field" would be constant in sky-light continuity, though not necessarily in brilliancy throughout the entire terrestrial sky, there would appear to be a diminution of sky-light concentration away from the "Milky Way" section.

To illustrate, we will assume that Des Moines, Iowa, and a certain adjacent sky area is the terrestrial "Milky Way" as observation is had from a celestial land position over Des Moines. The Des Moines sky area and a considerable sky area extending away from Des Moines would present to telescopic observation the terrestrial sky area of seemingly most abundant light accumulation. That accumulation would mean more points of light, but not brighter points.

Every observation beyond that established and more pronounced "Milky Way" sky-light accumulation would necessitate telescopic observation and photography at an increasing angle to facilitate search for "stars" on the distant horizons of the terrestrial "Heavens above." The detection of remote terrestrial "stars," or sky-light points, would find them more sharply defined as isolated entities than the sky-light accumulation comprising the so-called terrestrial "Milky Way." The brilliancy permitting of detection, of whatever intensity, or astronomical "magnitude," would accentuate the apparent isolation common to the sky-light of the entire Universe.

But that apparent isolation would not be as pronounced in the "Milky Way." The greater the volume of massed light, despite the lesser brilliancy of every point thereof, the less pronounced is the apparent isolation of each point of the entire area. However, the massed light-point whole constituting the "Milky Way" must appear to be more detached from other detected sky-light points of the entire sky. That is why the so-called "Milky Way" seems to be unique, yet it represents sky-light the same as any other detected lonely "star."

Though we would know from the celestial observation point that there existed a continuity of land and sky at the designated terrestrial "Milky Way," considerable of the sky-light area would not be detected as observation at an angle was made away from the Des Moines sky's center of the terrestrial "Milky Way." Any off-center observation imposes limitations. Though every terrestrial sky area is in fact to some degree luminous, as every area of the celestial sky is, many areas would have to be assumed nonexistent from celestial observation because the sky-light of such areas would not be detected for various reasons previously described.

The astronomical procedure of searching for "stars" on

ch. ends p. 145

the distant horizons beyond the "Milky Way" concentration of celestial sky-light may be considered co-related to the more realistic procedure of a laboratory technician's search. That realistic search would constitute examination of a mass specimen on the illuminated surface of a clinical glass slide. The multiple minute particles of the specimen mass would be the technician's field, as the entire celestial sky is the astronomer's field. The electric-light illumination of the glass slide would represent the astronomer's sky-light. The technician's microscope would represent the astronomer's telescope.

In direct and near-direct focus of the microscope lens the greatest accumulation of specimen would be apparent even though the field was of the same density throughout. If the field were enlarged by lens focus, there would have to appear to be a diminishing of the central concentration of specimen. Then the original margins of the central concentration would have to appear to become thinner, to a point of specimen obliteration. The development of that condition would not mean that there was actually less specimen substance at the extremities of the glass-slide field, but it would limit observation of the field equal in density. The area of direct or near-direct lens focus would seem to hold the most specimen substance.

It becomes evident that the laboratory technician, "working in these walls of time," holds a considerable advantage over the astronomer working in the limitless corridors of infinity. The technician working in a limited but realistic world can constantly move and adjust the glass slide, or "star field" equivalent, to serve his purpose. And he can keep constant, or he can increase or diminish, the illumination of his field. Further, in having complete control of the field and its light, he can at will adjust the microscope lens for constant dead-center observation of the specimen.

There seems to be lacking any record of an astronomer who was capable of making adjustments to his "star field" specimen which would keep it in direct focus, immobile, and under the constant and proper illumination required for observation and determination. Sky-light of the celestial, as well as the terrestrial, is not subject to the penetrative enterprise of telescope lenses or to the whim and deduction of astronomers. On the contrary, sky-light everywhere influences lens ability to detect as well as the astronomer's deduction. It is a fascinating game of tag, where the astronomers and their lenses continue to be "it."

The humble but much more practical laboratory technician holds an additional advantage, in that he or she deals with known entities in a world of reality. If the least doubt is harbored concerning the identity of certain matter or entities within the specimen of the slide field, any number of practical tests made directly upon the doubtful substance will determine its exact properties. That little feature of *direct contact with and immediate test of* the questionable entity differs considerably from the extremely abstract mathematical tests to which the astronomer is restricted in an effort to determine conditions and entities of his remote abstract "star fields." It will be shown that astronomy refutes astronomical conclusions in the making as a result of the manner of observation leading to the conclusions.

Where an astronomer detects dual movement, or what appears to be dual, in observation of a remote luminous celestial sky area, and spectroscopic analysis[3] confirms apparent duality of motion, he is compelled by concept to conclude that two distinct entities are operating at the single light point under analysis. The astronomer could, but he does not, conclude that a *single energy* at work at the particular celestial sky-light point is prescribing a *double motion.*

In consideration of the astronomer's conclusion, it is here pertinent to recall previous reference to the undulation

[3] Spectroscopy is the precise study of color as generalized from visible light to all bands of the electromagnetic spectrum.

ch. ends p. 145

motion of sky gas, and that the astronomer even makes use of the word "undulating." And it may be well to remind that undulation is a double motion.

The astronomer is forced to conclude that the motion is attributable to entities contained in the astronomer's mind. And the entities of illusion the mind contains are "isolated bodies," globular or spheroidal, moving in a circle or an ellipse. Nothing else will do. In reality, there exists for telescope lens and the astronomer's instruments to determine nothing more than the *dual motion of gas* in a luminous sky area which covers and obscures the stationary land under that detected sky area. The active sky gas moves, but the underlying land never participates in the movement.

It seems singular that the astronomer determines in favor of the preconceived "circling or ellipsing bodies" in view of the fact that he applies the very meaningful terms "moving back and forth," "undulating," and "fluctuating," which deny the preconceived entities and their motion. Yet his illusion-fostered conclusions must be that the lens and the spectrum, or either, in recording such movements truly establishes the existence of two distinct celestial "bodies" in motion.

To emphasize this most important feature, it should be noted that his conclusion of celestial "bodies" does not imply bodies of gas in keeping with the dictates of reality and reason. To him the illusion persists that the *motion of sky gases* signifies the motion of motionless land mass, which cannot be detected under the *luminous moving sky gas*.

Observe that nothing has detected or established even one mass body in motion, to say nothing of two bodies. There has simply been achieved confirmation of double motion, within a certain luminous celestial sky area. Hence the astronomer's terms "undulating" and "fluctuating" are appropriately applied for description of the recorded movements of gaseous elements within the luminous sky area. But the terms have no further application.

Upon that single instance of erroneous conclusion is erected an astronomical framework of abundant miscalculations. Having checked the mechanical findings of double motion with that found by direct vision, there is nothing left for the astronomer's conclusion than that which his concept holds: "isolated rounded bodies circling or ellipsing in space." The telescopic and photographic lenses have not detected and recorded them; the astronomer has not observed them. They, the "bodies," are not established by spectrum and spectroscopic analysis. However, they are concluded to exist as isolated globular mass entities, when they constitute nothing more than lens-created disk areas of sky-light gas in motion.

We may duplicate the astronomer's application and his findings of the celestial by returning to the lofty stratosphere observation point permitting view of terrestrial sky areas. As we adjust the telescope for observation of Portland and Bangor, Maine, on the east coast of the United States, or any other section of the nation, the luminous sky areas to be detected over any land community will appear precisely as the luminous celestial areas of astronomical observation appear. Our lenses will detect nothing but a luminous disk-like sky area. At every angle of observation and as far as our lens can penetrate, we will observe the same condition. It would be ridiculous even to hope to see through the luminous terrestrial sky areas to observe the land and water and the community life we know is underlying the sky areas.

We may first detect the sky-light over Bangor, Maine. It will be found that Bangor's sky-light seems to fluctuate. It will be prescribing the dual motion which could very readily be misinterpreted as "circling or ellipsing" from proper distance. Were we to achieve that distance, there would develop the illusion of circling. And though we might even accept the illusory movement as having application to the luminous sky

ch. ends p. 145

area, our knowledge of the underlying land would dispel the illusion in relation to the land area. We would not fleetingly harbor the illusion that Bangor had become isolated from the remainder of Maine and was executing an orbital waltz in stratosphere space.

Making telescope adjustment to embrace terrestrial sky areas north of Bangor, we may detect a luminous terrestrial sky area that appears to roll. And it will be much brighter than the "star" of Bangor. We will perhaps find on consulting our terrestrial "star chart" that the bright rolling area represents the sky over Montreal, Canada.

As we continue our telescopic search, there will be detected a luminous sky area west of Montreal which arouses interest. There will be a pronounced white film on the lower left corner of the sky area. Its appearance will promote doubt that it is part of the sky area, and we shall conclude that since it is not of the luminous sky area, it is a "nebula" in the stratosphere.

Then, adjusting our telescope for observation of the New Hampshire sky, we shall detect a dark area in the luminous sky which our "star chart" designates as Portsmouth, New Hampshire. Magnifying that luminous sky area with a stronger lens will disclose the original dark spot as three distinct formations. They will be easily considered humps on the luminous sky area. In fact, they will so closely resemble the astronomical "Camel Hump Cluster"[4] in celestial sky-light that we will be impelled to name them the "Triple Humps of Portsmouth."

Hence it will be perceived that the conditions recorded of luminous celestial sky areas, where light shading is at one time determined as a "nebula" detached from the luminous

---

[4] Some Arabic cultures perceived the stars of Cassiopeia as a camel: its head was composed of Lambda, Kappa, Iota, and Phi Andromedae; its hump was Beta Cassiopeiae; its body was the rest of Cassiopeia; and the legs were composed of stars in Perseus and Andromeda.

sky area and on other occasions as a grotesque formation of the luminous area, must be included in record of terrestrial sky areas. As it has been related, corresponding conditions have to date been found in the luminous terrestrial sky over White Sands, New Mexico, and adjacent territory. As the sands of this Earth's desert regions are related as particles of sand, and as the waters of the Earth are related as water, in like manner does the luminosity of every terrestrial sky area correspond to elements and conditions of celestial sky areas. Terrestrial sky gas describes the identical motions of celestial sky gas. And the observed conditions of terrestrial sky areas will impose the same illusions as those burdening astronomers' empty quest of the celestial universe about it. The identical "stellar spectra" will develop from analysis of light waves from terrestrial sky areas as presently developed of light movement in celestial sky areas.

Massive astronomical compilations of the centuries have unknowingly directed man's course away from observation and comprehension of the realistic universe about us. But the current opportunity to view terrestrial sky-light function and the ensuing formations abrogates[5] astronomical presentations. And that modern view eloquently attests to the import of ancient philosophical dictum: "On Earth as it is in Heaven."[6]

Modern enterprise confirms that what is to be found in the celestial "Heavens" has undeniable counterparts in the terrestrial "Heavens." And it has been vividly disclosed that it is the deceptive appearance of things and conditions over the land areas of the Universe, rather than that which exists on land under the celestial and terrestrial "Heavens," which has made for confusion, thus denying acquisition of the universe about us. The same astronomically recorded shifts in the spectrum, from the longest red wave to the shortest violet

[5] In this context: negates or invalidates.
[6] From The Lord's Prayer, Matthew 6:10.

ch. ends p. 145

wave, are to be registered from observation and analysis of terrestrial sky-light movement. The synonymity of celestial and terrestrial sky-light performance, meriting the same interpretation, must provide evidence for the least discerning person that astronomy's announced celestial values are purely illusory.

It may thereby be perceived that were we to apply the astronomical yardstick to the terrestrial sky's luminous outer surface, certain areas would, like the celestial area named Sirius, be assumed to possess more than twenty-six times the Sun's mathematical candle power. The absurd conclusion would develop from such terrestrial sky area's *apparent heat intensity*. We repeat, *apparent heat intensity*.

Fantastic? How could it be otherwise, with our physical knowledge of terrestrial sky areas? Yet, that would be the inevitable development when we attempted to gauge the terrestrial sky with the same instruments utilized by astronomy for gauging the celestial sky. In such application of astronomy's gauges to terrestrial sky areas, it will be established that the red and the green waves hold no such meaning as that which is astronomically concluded from celestial sky-light areas where the colors are evidenced. The tests to be made of terrestrial sky-light will establish the value of red and green waves from terrestrial sky-light to be diametrically opposed to astronomical deduction.

Ancient observation of the lights detected in the universe about us developed the so-called "star charts." That development was an artful expression of the wholesome "star" observing pastime. Nobody was deluded through the art of celestial light charting. But when the same art bedecks itself with the judicial garb of science and imposes upon the world illusory conditions acclaimed to be real, there is described neither art or science.

During the many centuries of observation, there should have been discernment of the illusions. And the least that might have been achieved was comprehension of the unfailing manner in which all creative energy must move. That movement is a wave. But the universally manifested wave motion was replaced by the astronomical fraternity with the barren guess of "circling" or "ellipsing." And, strangely, such replacement was made to sustain theory even as the wave term received empty lip service. With that replacement from the world of the illusory, the entire astronomical structure erected upon the "circling" or "ellipsing" guess becomes purposeless and void. Nowhere throughout the broad domain of research in pure and applied science is there to be experienced the "circling" or "ellipsing" motion contained in and making the foundation for celestial mechanics. Wherever such motion seems to take place, other than in man-made mechanics at terrestrial level, it is purely illusory.

With relation to the motion of universally dispensed energy, it is timely to relate a personal experience confirming that creative energy, wherever manifested, is compelled to move in a wave. That holds true even if every lens the world possesses causes the motion to appear as circling. The lens is incapable of faithful recording, but the brain should be aware of such fact; for it is the brain that truly sees.

In the chapter dealing with the pilgrimage, a meeting with the famous physicist, Dr. Robert Andrews Millikan, then President of the California Institute of Technology at Pasadena, was described.[7] At that time, during the summer of 1928, Dr. Millikan's able assistant was Dr. Carl Anderson.[8] And as Dr. Anderson conducted this then-youthful enthusiast over the institution's campus to view the world's first isolated electron, he remarked, "The electron prescribes a circling motion."

[7] See pp. 37-38
[8] Carl David Anderson (1905–1991) was an American particle physicist who shared the 1936 Nobel Prize in Physics with Austrian-American particle physicist Victor Francis Hess (1883–1964) for his discovery of the positron, the study of which began under the supervision of Millikan.

ch. ends p. 145

In manner lacking diplomatic nicety, we responded, "It does what, Dr. Anderson?"

Dr. Anderson replied, "It *seems* to move in a circling manner."

With the same lack of diplomacy, we answered, "That is better."

Though Dr. Anderson was a very learned physicist who was subsequently awarded the Nobel Prize, he referred to the electron's seeming motion even though his brain saw the true motion. Such mention of circling was due to the influence of the seeming motion. And the lens was responsible for that seeming condition.

Yet it was known to one who had never observed an electron that the basic and irrefutable principles of motion precluded any possibility that the electron performed any circling.

In the case of the mathematico-astronomer it is found that, despite knowledge of the wave and bend of energy, there is a persistent adherence to the seeming, or illusory, motion. His unswerving devotion to the illusory demands denial of the authentic motion in all astronomical observations and conclusions. Hence result the numerous miscalculations of that motion's distance and speed from the astronomical point of observation. And it precludes possibility for understanding of the heat engendered at the luminous celestial sky area where the motion is detected.

No structure in a world of reality can be sustained on a mythical foundation. The framework of astronomy is productive of nothing realistic, because it is erected on the illusory. Worse, the constantly increased lens magnifications of the luminosity projecting the original illusion retards findings of fact in the realistic Universe. Is it too much to expect that after three hundred years of mathematized telescopic astronomy, following three thousand and more years of astronomical art, the illusory framework must be discerned by government agencies. Their findings have uncovered the basic illusion and have paved the way for the astronomers' redetermination of cosmic values.

Though theory may be of enduring mathematical prescription, it is always subject to change. Along the course of civilization theory which represented the truth of each time and place has undergone change for the better. That process of change has made civilization. From the time of Hippocrates,[9] the science of medicine has been subjected to the most intent scrutiny by members who have dared to question its premise. And their questioning made for redetermination of anatomical values which benefited humanity and advanced medicine to its present high estate. It was only through persistent doubting, contradicting, and experimenting that factual knowledge was acquired of the human body's circulatory system. And with that redetermination of values a thousand and one progressive and helpful features were evolved. They could not have been possible until the false theory of blood function had been discarded.

To project the circulatory system of man into the arena of celestial sky-light analysis affords a timely comparison of values. It may serve to clarify features of Physical Continuity which the atomic physicists very nearly found with their determination: "There is a play of energy between particle and particle of the entire Universe."

For the past three hundred years mathematized telescopic astronomy has sought to determine the creative "circulatory system" of the Universe. But in that search it insisted that the universal blood flow — magnetic force and sky-light gas — was restricted in its function to the terrestrial side of the Universe body, or whole. Here the continuous and constantly energizing sky of the Universe whole is likened to the human

[9] Hippocrates (c. 460–377 BC) was a Greek physician and philosopher, traditionally referred to as the "Father of Medicine" in recognition of his lasting contributions to the field.

ch. ends p. 145

body's circulatory system. The sky veins function throughout the Universe body under the force of actively circulation sky gases. The gases are in turn constantly agitated, or stimulated, by the creative magnetic force of the Universe.

The terrestrial represents but one side of the Universe body. The celestial represents the other side. The creative forces at work do not nourish and stimulate one side to the neglect of the other. Were such the case, the terrestrial only could survive.

To judge from astronomical conclusions, neither universal magnetism nor celestial sky gas exists. And where they are reluctantly conjectured to exist, they are so misinterpreted and miscalculated as to obscure their function and purpose. The astronomer concludes that the formidable sky gas circulatory condition, which actuates the terrestrial and the celestial, is negative as a continuous vein of the Universe whole. Hence the abundant vein expressions, light variations, light shadings, and distortions are not considered developments of a sky vein extending through the celestial.

The determination that such celestial sky expressions are not from celestial sky gases, and the conclusion that many expressions are remote from the luminous celestial areas, has been responsible for the most complex system of contradictions within the history of all the sciences. In consideration of astronomical procedure, it is not be wondered[10] that such a conclusion should result as that *matter* existing in so-called celestial "nebulae" has density a million times *less* than anything on Earth. By such a figure so-called "nebulae" are astronomically ordained as *matter though less than matter*. The matter of reference is celestial *sky gas*, and it has identical terrestrial sky-matter, or gas, weight. Hence it is sky gas, which is not matter as commonly indicated by the word. But the astronomical conclusions present something

[10] The phrase "it is not be wondered" is an archaic way of saying "it is no wonder."

more sensational. They compare celestial *sky-gas weight* with terrestrial *land-mass weight*. The absurdity of comparison should be evident to a ten-year-old child.

In previous examples, particularly the white "cloud" accumulation in a photographed area of terrestrial sky-light, it is shown that the astronomical "nebula" is nothing more than moving gas of and within luminous outer sky areas of the celestial and the terrestrial. To accredit such gas "nebulae" with the weight of mass, as mass is considered in a world of reality, is equivalent to attributing mass property to an ectoplasmic emanation in the field of the spiritualist. Though it is true that even electric impulses have a certain weight, one would hardly consider comparing the relatively weightless electric impulses registered from brain-mass functioning with any known mass property.

On the opposite end of astronomy's mathematical seesaw, it is disclosed that some "stars" possess density a million times *greater* than anything found on Earth. Assuming that the astro-mathematicians, who through their own choice of words and figures prove that their estimates deal exclusively with sky-light and its expressions, could by some necromantic performance accord such weight beyond known mass to luminosity detected and analyzed, what meaning can it have in a world of reality? What can it mean to have an acre of land or a grove of trees a million times the known and *real density* of an acre of land or a grove of trees? The human mind cannot estimate established mass density. What would it do with a million times known density?

Hence a million times the density of known density can mean nothing more than a choice of words meaningful only in the unreal world of the astro-mathematician. Any attempt to apply to known density a million times its known characteristics as density transcends conceptional capacity. Moreover, multiplication by a million would necessarily abnegate density as known density and thereby would

ch. ends p. 145

establish density as *something else* beyond density. In the sequestered realm of hallucination, it might provide a nucleus for some heretofore unexpressed, or expressed but unrecorded, fantasy of confusion. Otherwise, it expresses only the multiplication which should be registered: a million times one million ciphers equals one million times one million of nothing etc., *ad infinitum*.

To clarify this material relating to mass property and gaseous content, it can be observed that there should be marked differentiation of the subjects. They cannot, in this instance, be considered interchangeable — although in final analysis, they may be considered interrelated:

1. Astronomy and its unlimited mathematical scope of operation can deal only with observation and deduction of the luminous celestial gaseous sky surface. "Surface" here means the luminous *outer sky layer* detected by the telescope lens or, if undetected, mathematically considered to exist.

2. Though there is limited "weight" to sky gas over celestial land areas, it holds no weight significance when compared with the underlying undetectable landmass weight. And the fact that celestial land cannot be detected by astronomy's artful instruments and measurements can bear repetition on every page of this book, for there reposes in that feature the basis for comprehension of the realistic Universe.

3. Yet it is found in astro-mathematical conclusions that the gaseous sky of some celestial areas possesses density a million times more than *anything* found on Earth. Were it to be concluded that celestial sky gases of some areas weigh a million times more than terrestrial sky gases, we could blame mathematics and forthwith

relegate the subject to Dante's Inferno[11] or some corresponding site. But painfully it is concluded that the sky gases are that much weightier than *anything found on Earth*. And unless words too have become subject to astro-mathematical magic, the astronomical conclusion means the land content of the Earth, not the gaseous sky content over the Earth.

4. Further the same astronomical methods disclose that so-called celestial "nebula" is of density a million times *less than anything on Earth*. Again it is found that there can be no comparison. Earth land mass and celestial sky gas are by no means the same or similar subjects.

The freely utilized infinite mathematics of Immanuel Kant[12] hold such absolute power over the astro-mathematician that they can endow such subjects as terrestrial land-mass and celestial sky-gas density with synomymity. Of such mathematical stuff are "stars" made. The constituting material may be a million times *heavier or lighter*.

It may be of value to observe:

(*a*) The light detected or deduced is from gas which represents a "star."

(*b*) The *moving shadows* in that gas are possessed of density a million times *less than anything found on Earth*.

(*c*) Then, elsewhere in the labyrinth of astronomical archives, it is unhesitatingly recorded that a certain other "nebula" possesses density thirty-five hundred million times *the Sun's mass*.

(*d*) In the last case it was noted that the ectoplasmic substanceless "nebula" is not assumed to weigh that many times

[11] A reference to *Inferno*, part one of three of the epic Italian poem *Divine Comedy* (1320) by Dante Alighieri (c. 1265–1321).

[12] Immanuel Kant (1724–1804) was a German philosopher and one of the central Enlightenment thinkers; his comprehensive and systematic works in epistemology, metaphysics, ethics, and aesthetics have made him one of the most influential figures in modern Western philosophy.

ch. ends p. 145

the Sun's *surface light mass,* it is assumed to weigh thirty-five million times the unknown *mass content* of the entire Sun.

Such an estimate of the Sun is postulated with impunity in spite of the fact that nobody has knowledge of the meaning of "Sun" other than that it gives light, heat, and energy. Hence how can there be an estimate of the mass weight of that which is unknown? Yet astro-mathematics will provide the weight estimate without knowledge of what is being weighed. Such is the power, but hardly the glory, of infinite mathematics.

It becomes increasingly evident that our earliest ancestors, who worshiped that Sun without the questionable benefits of modern astro-mathematics, knew more about the Sun than the modern mathematical astronomer does.

For a determination of values, it should here suffice to record that all such mathematized conditions of weight assumed at celestial level would have application to terrestrial areas under investigation from any part of the celestial. Though it is definitely known that such mathematized and assumed celestial conditions do *not* exist on terrestrial land areas or in luminous sky areas, they would have to be mathematically concluded to exist, if for no better reason than that of sustaining the doctrine "Figures do not lie." Though God forsake His kingdom and the Universe collapse, the figurative must prevail; the figure must never be questioned. For if there be no Universe, the figure will create one. And if there be no Creator or Creative Force, the figure will adequately replace it. So says the figurer.

Astronomy holds a unique, most unenviable position. It is unlike any fruitful science known to man. Its premise is eternal, though it be the most illusory ever established.

Philosophy, seeking to find behind things and events their laws and eternal relations, dares to abandon a premise found to be at variance with fact. Only in such manner can philosophy continue to seek for, determine, and interpret values in the world of reality. Though philosophy's broad horizons extend the things and conditions of the physical world into the metaphysical realm, there is ever a continuity of pattern wherein things and conditions for a physical plane continue to be reasonably identified on the metaphysical plane. But despite its broad scope, philosophy need not resort to figurative definition of its transcendent values. Obscuring equations and symbols are not required for coherent description of factual values interpretable by words. Where there is a fact to convey, words will be found to express it. But when there are no facts, mathematical symbols very formidably obscure the condition.

Astronomy, claiming to interpret the physical Universe, possesses knowledge of neither the beginning nor the end of its telescopic domain. Nor has that domain origin or ending in a world of reality. Sky gases misinterpreted as land mass can hardly be considered expressive of reality. Nor can the gross misinterpretation of energy's wave motion to be prescribing a "circling" or "ellipsing" motion assist man's comprehension of the created and realistic Universe and afford closer attunement with the infinite.

"The Heavens proclaim the glory of God."[13] And they would proclaim that glory if a telescope had never been invented. After centuries of telescopic astronomy, man beholds the same luminous splendor displayed for his earliest ancestors. He sees no more and he knows no more of the celestial "Heavens above."

Though telescopes have found more points of light for the telescopic lens, they continue to be incompetent to penetrate such light points and to permit determination of realistic value attaching to the lights and what is under the lights. Further, the abstract mathematical values imposed on lights detected have so distorted real created values that they have

[13] Psalm 19:1

ch. ends p. 145

become progressively more obscure with each advancing year of telescopic detection and astronomical interpretation. In fact, the abstract mathematicians have so mathematized the real Universe that it has been made a figurative Universe where only mathematical symbols may dwell.

Therefore, one can both mentally and physically indulge the real Universe through understanding of the importance of current events. Then can one fully benefit from the creative splendor of celestial sky-light, despite the obscuring and distorting astro-mathematical conclusions resulting from basic fallacy representing astronomy's Prima Causa.[14]

Timely understanding of cosmic values recently discovered enable one to discern why a great churchman, the late Cardinal William O'Connell, Archbishop of Boston, publicly denounced the atheistic tendencies of abstruse mathematics in the summer of 1927. At that time, His Eminence confided, "Science is going around in circles." The unprecedented events of our time, as here recorded, eloquently attest that if the phrase "going around in circles" ever merited application it could have no better application than to that abstract science of astrophysics that the cardinal had in mind.

The cardinal's timely observation was subsequently amplified by the late Garrett P. Serviss,[15] who wrote of the author of that "beneficent" mathematical postulate: "As concerns the intellect of the average person he is responsible for having let loose from their caves a bevy of blind bats whose wild circling in the limelight of publicity draws dreary gleams around the moorland of everyday commonsense."

Where is the meaning in mathematical gymnastics providing a presumptive estimate of our Sun's weight one billion or ten billion years in the past? The meaning is less, if there

---

[14] Translated from Latin: first cause; a phrase that refers to the initial, ultimate cause from which all other causes and effects originate.

[15] Garrett Putnam Serviss (1851–1929) was an American astronomer and early science fiction writer.

could be less meaning, when other mathematical dictums contradict the estimate and establish that the Sun's realistic magnitude and function is unknown.

What meaning to "the life of a 'star'" and its mathematized weight? And if every word of that question had application to a world of reality, what would it contribute toward man's comprehension and acquisition of the universe about us?

What value to the astronomical estimates of thirty thousand million, two hundred thousand million, and five hundred million celestial light points, when the meaning of just one point of light is not understood, at least not by the astronomer?

No physical science could or would accept for three weeks, to say nothing of three centuries, the illusions of astronomy. The physical sciences could and would determine the reality of premise before elaborating on the premise. But what could astronomy do? The astronomer's powerful mathematical conveyor could not take him to the celestial sky-light points under investigation.

In geology, biology, physics, chemistry, anatomy, botany, the findings are substantially rooted in the world of reality. And though at times figures are applied in such truly scientific endeavor, they have basis in reality rather than in illusion. They are intended to enlarge but never to distort the basic reality, and the mathematical results, though always subject to direct and most critical scrutiny by brain sight rather than lens sight, are immediately questioned, and as readily rejected, if they are at variance with fact.

Within the broad scope of positive and applied sciences, where the formula for duplication of man is unknown, the fact is freely admitted. Abstruse figures are not paraded to assume the laboratory making of a real human being or to facilitate the deception of having made a super Frankenstein monster to replace man.

What value could possibly attach to the mathematical

ch. ends next p.

making of a single drop of blood which the combined sciences are unable to reproduce in laboratories of a world of reality? In spite of the mathematical formula, the Red Cross would be obliged to continue the more realistic practice of extracting blood from the veins where Creative Force caused it to be installed and where only Nature, agile agent of that Force, is capable of reproducing it. Would the most precise and positive dictums of Immanuel Kant's infinite mathematics *actually provide* a single drop of blood? As concerns a world of reality, infinite mathematics are as nebulous as infinite space.

Contrary to all scientific endeavor and conclusions within an established order of reality, the mathematical astronomer is privileged to create mathematized entities having no relation to the world and the order of reality. Further, he is permitted to distort and obscure entities abiding in a world of reality through the play of abstruse mathematics.

A most important aspect of that world of reality is the sky which envelops the world's land and water, vegetation and life. And its luminous outer surface mystifies men with unique performances against the dark curtain of infinity's stage. It presents the most intriguing spectacle in the Eternal Theater owned by that unknown Peerless Producer of celestial and terrestrial drama. That magnificent Universe Producer endowed the most remote celestial area with the identical physical values common to this known terrestrial area where we dwell.

And in the creative course of such transcendent production, there was also evolved the brain of man. The Producer intended it as a formidable agent to check and correct *the illusions developed* form man's feeble observation of the creative production. Every celestial mile of that production known as the Universe is as realistic as this Earth area is. And it is denied such created realism only as a result of terrestrial man's faulty observation and faultier interpretation. Where the Producer intended the brain to see truly, man isolates the brain and delegates its duties to the lens. It doesn't work.

Therefore, the roads of illusion are everywhere. As they have been proven to exist through actual photographs over the luminous terrestrial sky areas of White Sands, New York City, and elsewhere, they extend over every luminous sky area of the entire Universe. There is not a mile of that celestial area described by the astronomer's so-called "star" chart, or factual sky chart, which does not present the identical road of illusions to be encountered in every journey over the illusion-producing luminous outer sky areas of our Earth.

Since that claim was first made in the year 1927, the stratosphere ascents and the lengthy series of U.S. Naval Research Bureau rocket flights have procured photographs of luminous and deceptively isolated globular terrestrial sky areas confirming the claim beyond a question of doubt.

"With eyes ye see not, yet believe what ye see not."

8
Into the Unknown

"The greater the knowledge, the keener the pain."

Though the world's dreamers are sufficiently endowed with knowledge of a transcendent order, they are denied knowledge of the price their dreams will exact. Perhaps it is well that such is the case; otherwise the world might never learn of the dreams.

As the dreamer of 1926–27 could not foresee the flagellation[1] his dream would inflict, neither could he anticipate the stupendous forces to be mustered for his dream's confirmation. It was almost twenty years to a day, in October, 1946, when the most powerful force for confirmation began to function beyond his most ardent expectations. It brought realization of his hopes of twenty years before, when he had visited another of the world's pioneering eccentrics in the person of Dr. Robert Goddard at Clark University at Worcester, Massachusetts. Dr. Goddard was then painstakingly experimenting with rocket construction in his cell-like laboratory at the university. He too was denied funds for the perfection of his particular dream. And he heard the customary mockery reserved for dreamers of all ages.

[1] In this context: scourge; suffering.

Though there was then realized the possibilities of Physical Continuity's confirmation through the medium of the rocket, there was little expectation of the rocket's early perfection and the extraordinary part it was destined to play in procuring confirming data. Hence there was unrestrained enthusiasm when, in October 1946, the U.S. Naval Research Bureau's V-2 rocket was sensationally projected into the perpetual stratosphere darkness beyond the sky enveloping the desert community of White Sands, New Mexico. There, at the altitude of sixty-five miles, its camera developed from the terrestrial sky area being photographed an undeniable replica of that which had been described as early as 1927.

That original photograph over White Sands conformed in almost every respect with the revolutionary drawing of 1930. The only difference was that the rocket's drift developed an angle view of the disk areas presented by the drawing. Had the photograph been on the perpendicular, there would have been developed one of the drawing's luminous disk areas. That original 1930 drawing of terrestrial sky-light illusions has been reproduced as **Figure 4**. It merits reader observation and study because it is the key for realizing factual Universe values.

The U.S. Navy's rocket camera photographs proved that any camera lens at sufficient stratosphere altitude will show every photographed outer sky area of the Earth as a luminous and deceptively globular and isolated entity, or "body." The photograph contains an angle view of the disk; a photograph on the perpendicular would show one of the assumed "isolated bodies" telescopically observed of the celestial. It proved the illusion in centuries of astronomical observation of the universe about us, for the luminous disk surface area must impose the delusion of an isolated globular "body."

In the light of such sensational rocket-camera performance within infinity's dark stratosphere corridor, high hope was held for the photograph's influence. It was reasonably

ch. ends p. 160

believed that the photograph would arouse the lethargic guardians of the mathematical Universe and afford realization of the sky-light illusions of the ages. However, in spite of such memorable achievement, there was no apparent awakening of the self-appointed arbiters of the Universe pattern. Their evident lack of discernment accentuated the Christly dictum: "None are so blind as they who will not see."

Accordingly, even as the remorseless truth of previous unorthodox disclosure was presented, the globular misconception caused the development of a series of misinterpretations of that photograph and others that followed. The misinterpretations represent forlorn attempts to keep intact the fallacious mental portrait of a mathematically isolated globe Earth. Though stratosphere photographs of terrestrial outer sky areas hold abundant proof that globularity and isolation are illusory, their message is too profound for understanding and acceptance.

"My truth is the truth." So say we all. It is sacred, and it must be preserved, even though it contradicts fact. Hence to escape the reality which would dethrone the accepted truth, the terrestrial sky-area photograph at sixty-five miles was concluded to be an area of the distant celestial. That conclusion, though lacking foundation, stemmed from the assumption that the rocket camera had tilted as the rocket, reaching its flight limit in the stratosphere, turned and began its descent, and the first photograph was assumed to be a segment of a celestial "globe body" millions of miles away. The fact remains that the camera need not have tilted, as assumed. The mere turning of the rocket in its gliding, or drifting, descent would have caused the camera to record at an angle the globular *terrestrial* sky area which the rocket was approaching. Subsequent photographs over the same terrestrial sky area confirmed the latter conclusion.

It is readily perceived that in the rocket's turning the camera lens could not reproduce the entire terrestrial sky area as it would have been photographed on the perpendicular. Hence at the second of rocket turning only an arc of the completed disk sky area could be detected by the lens. It resulted in an incomplete disk area being shown.* The camera lens' function was not changed. It was developing a disk through detection at an angle. Thereafter, it was compelled to produce only angles of a disk because the rocket continued to *drift*. There was no chance for a perpendicular photograph of the sky area. Had there been, the photographs after rocket turning would have shown a complete disk area comparable to those of **Figure 4**. Naturally, when any one of such disk areas is detected, it must deceptively appear to be isolated. There must appear to be space between the disk sky areas. That is what provides the basis for the isolation misconception.

The lens that was capable of converging luminous terrestrial sky area at a distance of fifty-five miles was therefore assumed to have photographed a celestial area assumed to be millions of miles away. Very interesting.

To avoid any possibility of confusion, let us assert that the figure fifty-five miles is accurate. Though the rocket's altitude was sixty-five miles, it was only fifty-five miles from the outer sky surface being photographed. The distance from the Earth's surface to the sky is from seven to ten miles; the ten-mile figure is utilized here for convenience, and the difference between seven and ten miles has little or no meaning for the illustration.

The lens detecting what was falsely claimed to be an area of the celestial produced an identical outline in subsequent undisputed photographs of the same terrestrial sky area from a distance of ninety miles. (The rocket's altitude was one hundred miles.)

It is to be observed that if the camera had been in the

---

* One should not confuse such a view of a completed disk with Professor Piccard's earlier photograph, which held an incomplete upturned disk.

ch. ends p. 160

rocket's tail, rather than in the nose, there would have been numerous full-disk photographs taken from the outer sky surface to the ninety-mile stratosphere flight limit. They would have been produced prior to the displayed angle photograph taken at the time of rocket turn in the stratosphere. After the turn, all terrestrial sky photographs have to be taken by a camera in the rocket's nose as the rocket descends in a long glide, or drift. They would show disk angles depending upon the angle of rocket drift during descent. The angle photographs would continue to be taken until the rocket again penetrated the Earth's outer sky on its return to land surface. Such was in fact the procedure in the original photographing expedition. Hence the photographs showing only an angle of the terrestrial are as they should be.

Moreover, though such an angle photograph need not have been of the immediate terrestrial sky area where the flight originated, it *would then have to be* a photograph of another terrestrial sky area beyond the point of flight origin at White Sands. Nobody has ever beheld a telescopic photograph of any celestial area presented as only an angle view of a disk or as a segment of one of the many millions of so called "globe bodies." The reason is that the astronomer's telescopes are firmly anchored. They are not drifting through space as the rocket-camera lens was doing when it detected luminous areas of the terrestrial sky.

Hence telescopic photography shows every area a complete disk. The ancient Galileo Galilei would not like only angles of a globe. He "saw" completely rounded "globe bodies," and completely rounded "globe bodies" they must be. And they are — but in the illusory.

The manifest contradictions ensuing from publicized accounts and copies of the terrestrial sky photographs were evidently not considered sufficiently misleading. There was presented for a popular mental journey in the circuitous land of assumption that which follows. A dark, aqueous-appearing area in the lower left-hand corner of one of the terrestrial sky-area photographs was proclaimed to be the Gulf of Mexico. There was, however, no mention of a light-penetrating medium being used. There are no doubt many who have enjoyed reading the interesting novel titled *Islands in the Sky*.[2] That title is in order for a book in the world of reality; but the designation "Gulf of Mexico in the sky" is another thing, not of the world of reality, since it is not a book title. The former deals with the world of reality. Books and titles are of that world, whereas the latter deals, and only inasmuch as any dealing may be had, with things and conditions in a world that is not.

To explain further, it is shown, that the photographs taken at an altitude of one hundred miles from the Earth's surface, or at about ninety miles from the terrestrial sky area being detected by the camera lens, had to present one of two things. Both conditions could not have simultaneously existed at the same terrestrial sky area. Either (1) the photograph with the aqueous-appearing area is a true photograph of an area of the Earth's surface, accomplished through the medium of infrared and extra-sensitive film which permitted the camera lens to penetrate the sky luminosity and reproduce the land surface under that sky area, in which case the surface details would not be reproduced with clarity; or (2) the photograph was not taken with infrared light, in which case the lens did not penetrate the luminous outer sky and the photograph does not portray water, as claimed.

Therefore, the area appearing as water represents nothing more than light variations and shadings of and within the photographed terrestrial sky-light areas. It is just another light-shading illusion like those developed in photographing celestial light. That light's natural activity has created and

[2] A 1952 young adult science fiction novel by English author Arthur C. Clarke (1917–2008).

ch. ends p. 160

continues to create many of the grotesque entities of the astronomical world.

To affirm suspected absence of infrared, there was not the customary mention of its application. If it was not utilized, the photograph's description has to be erroneous and expresses only that which was expected rather than that which the photograph contains. It is notorious that we all see only what we want to see, and believe only what we want to believe. It is truly held that "primed observations are as dubious as spies", the matter of the "Gulf of Mexico in the sky" seems to be a case in point.

The most substantial evidence indicating that the water-appearing area of the photograph is nothing more than light shading within a luminous terrestrial sky area lies in the fact that *the area did look like water*. The rocket-camera lens could not have penetrated through sky-light density without the aid of a special photographing emulsion, and if that emulsion was used it would have bleached the dark water under the luminous sky. It would have caused the dark water-appearing area of the photograph to be white, and therefore unlike a body of water in appearance.

Moreover, the Gulf of Mexico could not have possibly reflected its known physical characteristics under photography through light and at the recorded distance. Rivers photographed in aerial photography at altitude not exceeding five miles lose their physical characteristics as rivers and become mere lines, or streaks, on the land surface. Such a condition develops in photography which is not through sky-light. Hence, when the photographing distance is multiplied fifteen times and the lens is compelled to penetrate through sky-light with the aid of infrared, one could hardly expect a clearer portrait of the real physical conditions or objects being photographed.

Finally, by what favor of necromancy could a camera lens ninety miles from the photographed outer sky surface cause to be reproduced on the photograph the ninety-mile sky level and the one-hundred-mile land surface level? Particularly when one level was luminous and involved photography against the dark stratosphere background, whereas, the other level required light for a photographing background? And how could the developed photograph of both levels show that the entire photograph area was luminous except for the small dark area of so-called Gulf of Mexico water?

It would have to be concluded that there is no sky over the Gulf of Mexico. There was sky over the land area, because none of the land was shown. Had the lens penetrated the sky-light it would have detected land as well as water, but the so-called water area was but a small part of the complete photograph. Such modern magic would permit photographing the rug in one's living room and have an area of the developed photograph show a tub of water in a corner of the cellar while the remainder of the photograph showed objects in the living room over the cellar. Such photographic magic would be superior to the X ray, which in photographing one level seems to miss the other. In this comparison, the interior and the exterior become equal to photographing levels.

The simplest experiments establish that it is impossible to see what is on the opposite side of any luminous area or object. Try to look *through* the flame of a fire anywhere. Try to penetrate the luminosity of any kind of burner. It will be found that the luminosity of an electric-light filament, or even the feeble flames of a burning gas jet or of a common match, will defy lens penetration.

One must never lose sight of the fact that there exists no observing instrument that was not patterned after the human lens. The human lens is great and magnificent; but it is subject to many errors. Therefore, it must be held in mind that every lens holds the same elementary error as the optic lens. It demonstrates gross misunderstanding to claim that though the human lens is subject to error, the

ch. ends p. 160

photographic lens overcomes the inherent error. It does no such thing. If it did, there would not be curves developed by the photographing lens.

The advancement of telescopy through photographic recording of telescopically detected luminous celestial sky areas does not advance telescopic findings beyond the point attained when Galileo fashioned his telescope. At least insofar as the finding deal with the reality of celestial things and conditions, there has been no advancement. The mind of the astronomer must be influenced by the inherent error of the photographing lens as it is by the error of the telescope lens. And the enlargement of lens power in no way eliminates the error; in fact, magnification broadens the field of application for the original error. The unreal entities of such dual agents of detection are multiplied. And though the entities are unreal, they are more readily accorded the status of reality as a result of misplaced confidence in the ability of two detecting agents instead of one.

As one proceeds along the astro-mathematical lane of enchantment, one finds that a subsequent rocket-camera photograph, at an altitude of one hundred and fifty miles, contains white cloud-like formations. They appeared on the same plane as the remainder of the photographed luminous sky area. Strange to relate, as the dark area of the previously described one-hundred-mile-altitude photograph was misinterpreted as water on the land level ten miles under the photographed sky area, the white *light* formations of the new photograph were deduced as *clouds in the stratosphere above* the photographed sky area. Of course the white sky-light formations represent no such thing as "clouds in the stratosphere." All light photographs as white. And the white outstanding on the photograph was intensification of natural sky-light. The white light was more pronounced against the dark light shadings of a part of the photograph; hence though the white was more representative of sky-light, it was considered to be detached from the sky-light area. It was simply an aspect of the luminous terrestrial sky.

Lack of reasonable reference to *gas clouds* formed within that particular gaseous sky area recalls the apt announcement of a famous scientist: "The world of the mathematician is peopled by all sorts of entities that never did or never could exist on land or sea or in the universe about us." And we here take the liberty to add fittingly . . . nor in the luminous sky areas anywhere.

It may be appropriate to record that the clouds of common reference are restricted to formation within the Earth's region of atmospheric density. That region extends from sea level to about six miles above the Earth's surface. Clouds are produced as a result of atmospheric conditions prevailing throughout that atmospheric area. That same atmospheric region extends throughout the entire Universe, contrary to the conclusions of astrophysics. It need not come as revelation to stipulate that clouds, as commonly referred to in a world of reality, are supposed to contain moisture or the chemical potential for moisture. The moisture of such atmospheric clouds may develop into rain, hail, or snow. It would be extremely fascinating to witness the production of rain and snow from the gaseous elements of any sky-light area, where, because of the particular sky-gas elements, clouds could never form.

Celestial and terrestrial sky areas do contain *gas* clouds. But it would be a revelation if they were afforded due consideration in astronomical conclusions about celestial sky-light areas. That consideration would dispel a great deal of cosmic mystery and would permit even astronomers a view of the realistic Universe.

It would prove equally sensational to witness rain and snow from the stratosphere. If one harbors the idea that atmospheric cloud formation could develop in the stratosphere region of negligible atmospheric density, the thought

ch. ends p. 160

may be dispelled with knowledge of the factor denying stratosphere cloud formation. That factor is the cosmic-ray activity prevalent in the stratosphere at all times. Its forceful movement is ceaseless.

Hence insufficiency of atmospheric density and the constant movement of powerful cosmic rays prohibit cloud formation. The rays would ruthlessly rupture embryo cloud elements attempting to collect in the stratosphere. A stratosphere explorer described cosmic-ray activity as follows: "They bombarded the stratosphere gondola from all directions."[3] And if their activity could bombard a metal gondola, how much more effective would be their activity against a cloud formation?

Therefore, the problem raised by the announcement of clouds in the stratosphere over New Mexico is comparable to the negative problem of early scholastic hours when the problem presented denied the problem: "What happens when an immovable object meets an irresistible force?" Without the necessity for applying abstruse mathematics, it is to be discerned that an immovable object could not be known in the presence of an irresistible force, and vice versa. One must deny the existence of the other at the same time and place. If the object be immovable, it can experience no irresistible force, if the force be irresistible, there cannot exist an immovable object for that force. Hence for clouds, as commonly known, to exist in the stratosphere they would have to be more formidable as a force than the perpetual Cosmic Force behind cosmic-ray activity. That force behind is another seeming problem compounded by deduction.

Stratosphere explorers have experienced the action of cosmic rays, but there is no record of their having experienced clouds. An important aspect of the Copernican Theory was that the stratosphere, then unknown and unexplored, is

[3] Likely referring to the ascents of Albert Stevens since *Explorer I* and *II* were often referred to as "stratosphere gondola."

a vacuum, or an approximation thereto, where even cosmic rays have to be excluded for perfection of theory. However, the mechanical devices of modern stratosphere ascension and rocket flights have determined the presence and have registered the activity of heretofore unknown stratosphere elements. It has thereby been established that the early ether theory,[4] or conceptional void,[5] is only of assumptive value to sustain other assumptions of the theory.

The function of natural law, when the Universe was created, precluded any possibility of vacuum throughout the constructed Universe whole. And Nature, because of her perennial productivity, abhors a vacuum. She has nothing to work with in vacuums. The nearest approach to vacuum has been achieved by man in his terrestrial laboratories, rather than by Nature acting as a tireless agent of Creative Force throughout the Universe.

Therefore, in consideration of values established in a world of reality, the conclusion must be that the stratosphere photographs of terrestrial sky areas reproduce sky-light conditions exclusively. The dark shading is as much a part of the luminous sky area as the white. Such conditions correspond with those observed in luminous celestial areas.

And they establish that all necessary confirmation of the 1927 disclosures have been procured. Inasmuch as the photographs proved that terrestrial sky areas preset the same luminous and deceptively globular and isolated appearance as all other areas of the Universe, it is shown that every luminous celestial area holds the same chemical elements responsible for terrestrial sky luminosity. Hence the sky is universal. Since it is therefore established that the continuous terrestrial sky will deceptively appear to be comprised

[4] Aether (ether) theories propose the existence of a space-filling substance or field as a transmission medium for the propagation of electromagnetic or gravitational forces.

[5] Inversely, void or vacuum theories propose an empty space devoid of substance or matter.

ch. ends p. 160

of isolated globular areas, logic dictates that every seemingly globular and isolated area of the celestial is in fact as continuous and connected as the luminous terrestrial sky. That areas of the Earth's outer luminous sky deceptively appear globular and isolated makes it manifest that the globularity and isolation of celestial areas is likewise purely illusory.

Since there exists ample sky illumination to obscure the land at an altitude of ten miles, there is no possibility for rocket-camera lenses to penetrate the greater luminosity of sky areas at altitudes from sixty-five miles to one hundred and fifty miles. Photographs at such greater altitudes have a darker stratosphere background than at the ten-mile altitude. Hence sky luminosity is more pronounced and it represents a more formidable barrier for lens penetration.

To return to the period 1931–1935, the pioneer stratosphere explorer Auguste Piccard was unable to photograph any of the Earth's surface at the altitude of ten miles. That altitude permitted penetration only to the outer sky surface. However, though Piccard had not emerged into the stratosphere proper, his publicized description of what he saw was, "The Earth appeared as an illuminated upturned disk."

The conclusion is sustained by Piccard's observation after the ascension of 1931: "The Earth was taking on a copper-colored tinge." That tinge represented primary illumination, it was sufficient to obscure the land only ten miles away. At the photographing altitudes of the rocket camera, the sky area had long since developed from the primary copper-colored stage into an extremely luminous seemingly globular area. As the fuller luminosity of the sky area was being developed because of increased altitude, the camera lens was drawing the sky area's partial disk into a complete and apparently isolated disk, so that the partial disk detected at ten miles was a complete disk, or "globe," at the greater altitudes.

No amount of increased lens power in the rocket camera could have altered the related development. In fact, any notable increase of lens power when photographing such luminous terrestrial and celestial sky areas will contribute to greater distortion of the luminous area and will in no way contribute to penetration of the luminosity. Increase of lens power will impose an oppressive magnification of the light and it will cause the light, which normally photographs as white, to present *a pockmarked appearance of light pits and fissures*. Then the sky-light area might appear to be covered with "canyons" corresponding to the so-called "canyons" shown in photographs of the Moon.*

As the optic lens projects the desert mirage to play upon one's fancy, the camera lens that developed light variations and light shadings in a luminous sky area over White Sands produces corresponding illusions which foster popular delusions of the universe about us. That lens is capable of projecting a lake or a canyon in the luminous outer sky over the lake-less and canyon-less Times Square land area of New York City, or in any other sky area of the Universe. The formidable factor of light distortion will cause the weaving of fantastic canyons in the luminous outer sky over the flat Sahara Desert and the equally flat wheatfields of Kansas. It has woven them in the luminous celestial sky enveloping that part of the Universe designated as Mars. The "canyons" of Mars have no more reality than that which would attach to canyons on the Sahara Desert and on the flat wheatfields of Kansas. Only as such "canyons" might exist on the flat unbroken plains and deserts of terrestrial reality do they exist for telescopic detection anywhere in the celestial. They

* The too frequently publicized astronomical "canyons on the Moon" and "canyons on Mars" are produced through the same agency of the illusory, the magnification and distortion of sky-light. Recently exhibited photographs of luminous celestial areas detected by the two-hundred-inch telescope lens afford eloquent expression of the distortions ensuing from magnification of luminous sky areas. The flaunted power of that lens, often referred to as "the white elephant of Mount Palomar," would create the same distortions in terrestrial sky areas if it were located on any celestial land area.

ch. ends next p.

are restricted to *the light of the sky*, and they are a natural development of the magnification of sky-gas movement.

As previously explained (and like the proclamation of an ardent wooer's love, it cannot be too often repeated), every area of the universe about us possesses the identical sky which covers the Earth. It is of varying shades of blue when observed from terrestrial and celestial land surface, and it is luminous when observed against stratosphere darkness. It should not be too arduous an effort to discern that every astronomically defined "star," "planet," and "nebula" is representative of celestial sky-light. There are many millions of luminous celestial areas that must deceptively appear to be isolated as "stars." The natural function of sky gas makes every area a potential projector of grotesque entities that never did and never can exist in a Universe reality.

Though there exists on every part of the continuous celestial terrain the physical characteristics of terrestrial territory — the plains, the mountains, the oceans, the rivers, and the lakes — no lens, regardless of its power, has ever detected such physical characteristics through the luminous sky. The intensity of sky luminosity has no bearing whatever on the power of the lens to penetrate it: the most brilliant light and the most vague light provide equal barriers to lens penetration.

Our modern ability to penetrate into the great unknown provides uplifting knowledge that the Creative Scheme does not conform to astronomical interpretation. The grotesque entities of astronomical definition are shown to be products of lens manufacture. Their value is mythical in the realistic Universe structure.

9
2,000 Miles Over Land Beyond the North Pole

Land of eternal darkness,
Fearsome and unknown,
Long hidden by theory and guess,
Your mystery now has flown.

"I'd like to see that land beyond the Pole. That area
beyond the Pole is the center of the great unknown."
—Rear Admiral Richard E. Byrd, February 1947

The United States Navy's polar exploratory force was preparing to embark upon one of the most memorable adventures in world history. Under the command of Rear Admiral Richard Evelyn Byrd, U.S.N., it was to penetrate into the land extending beyond the North Pole supposed end of the Earth. And it was sensationally to culminate more than four hundred years of vague conjecture concerning the Earth's northern extent.

As the hour approached for air journey into the land beyond, Admiral Byrd transmitted from the Arctic base a radio announcement of his purpose, but the announcement was so astonishing that its import[1] was lost to millions who

---

[1] Importance; significance.

avidly read it in press headlines throughout the world. That announcement of February 1947 conveyed in no uncertain terms immediate fulfillment of man's cherished hope to penetrate into land area of the universe about us. It promised appeasement of man's hunger for knowledge of a route into the luminous celestial mansions. And it promised that knowledge at once, not a hundred or a thousand years in the future.

Like every great truth, the simple truth of that 1947 announcement was not to be discerned. The announcement's lack of ambiguity in describing the celestial route rendered it, like the descriptive message of twenty years before, a truth stranger than fiction. And in a world of theory's fiction, who can be expected to credit that truth which is compelled to make its debut in garments stranger than those which attired the acceptable fiction of theory?

The words of message were momentous: "I'd like to see that *land beyond* the Pole." There was nothing complex in that expressive statement of fact, yet despite its simplicity, the statement had to be misunderstood by the many who, unlike the admiral from Boston, feared the unknown. The simple announcement provided such impact on popular misconception that it was at once distorted so that it might fit into the established fiction: there can be no land beyond the Pole, the admiral cannot possibly be going where he clearly states he is going.

Carefully note the remainder of the announcement: "That area *beyond* the Pole is *the center of the great unknown.*" How could the admiral have had reference to any mathematically established and then currently known area of the assumed "globe" Earth as prescribed by the theory of 1543? It must be conceded that the land beyond to which Admiral Byrd referred had to be land beyond and out of bounds of theoretic Earth extent. Had it been considered part of the mathematized Earth it would not have been referred to as "center of the great unknown." Were it part of the recognized "globe" Earth it *would be known,* not unknown.

To confirm the import of Admiral Byrd's announcement, one has only to examine the globe, which is symbolic of the Earth concept imposed in 1543. Try to find any area of land, water, or ice which encroaches upon the North Pole and which is not known. It will be seen that terrestrial areas extending toward the Pole from the East, from the West, and from the South are now very well-known and have been definitely established as terrestrial areas for many years. Is Spitzbergen or Siberia unknown? And do any such land areas extend *north beyond the North Pole?* They certainly do not.

It will be observed, however, that there is no land area denoted as extending *north from* the North Pole point, or *extending to the North Pole point from out of the North.* How could any land be shown, despite its now proven reality, under the terms of theory prohibiting the land's existence?

Hence the land mentioned by Admiral Byrd must lie *due north* from the North Pole. Therefore, it is within the conceptional absolute space that has been *assumed* to exist beyond given points north and south to sustain the globe-Earth theory of 1543.

If advanced moderns fear to relinquish the globe-Earth fiction, visualization of the land's location may be had through the simple process of adding another terrestrial globe at the northern extremity, or exact north Pole point, of the presently conceived "globe" Earth. Give that added globe the same Earth diameter, or length, or give it twice or one hundred times the terrestrial length. If it is provided the greater length, that will spare the tedious operation of adding more "globes" eight thousand miles in extent. The added globe will of course extend into space. Where else could it extend? The created Universe whole extends in the space where the Universe was ordained. As it is necessary to have relative land space to build a house, it was necessary to have absolute space to build the Universe.

Such is that land's location. It is not on the so-called "other side" of the Earth. We know both sides. It is beyond the point north where the Earth was *assumed* to end. It is endless in its extent toward and into celestial land areas under the luminous points observed "up," or out, from the known Earth area of theory.

In review of the magnificent naval accomplishment of February 1947, it is perceived that Admiral Byrd was not content merely to announce his desire to "see that land beyond the Pole"; but he did in fact *go beyond*, where he acquired observational knowledge of the physical aspects of that land he had referred to as "the center of the great unknown." Unlike the flight of fancy indulged in by the Boston cardinal and the early pilgrim of 1927, the admiral and his airplane crew accomplished a physical flight of seven hours' duration in a northerly direction beyond the North Pole. Every mile and every minute of that journey beyond was over ice, water, or land that no explorer had seen. (It is known that Raoul Amundsen, Umberto Nobile, and other earlier explorers may have witnessed conditions at the exact North Pole point, but they definitely did not see and travel over the land, and mountains, and fresh water lakes extending *beyond* the Pole and *beyond* the Earth of theory.)

The admiral's airplane pursued a course on the horizontal from the North Pole point to a point 1,700 miles beyond the Earth. Then the course was retraced to the Arctic base. At no time did he "shoot up," or out, from the Earth level. As progress was made beyond the Pole point, there was observed directly under the plane's course iceless land and lakes, and mountains where foliage was abundant. Moreover, a brief newspaper account of the flight held that a member of the admiral's crew had observed a monstrous greenish-hued animal moving through the underbrush of that land beyond the Pole.[2] In view of the popular misconception that it is necessary to "go through space" in order to progress beyond the Earth, it seems fitting to emphasize that there was land or water directly under the admiral's plane in his flight beyond the Earth. The land and the water were of the same consistency as land and water comprising this terrestrial area. There was nothing mysterious about the terrain. The atmospheric density provided oxygen content common to Temperate Zone areas of the known Earth. Above the airplane stretched the continuous sky; beneath it reposed the land. What more could one have asked of that which for many centuries had been conjectured to be "empty space?"

The magnitude of that memorable flight beyond the Earth but always over realistic land and water was never submitted for popular consumption. Press representatives were denied knowledge of it except during the brief period of active flight, when radio dispatches kept them informed. And insofar as personal knowledge extends, the admiral contrary to precedent, failed to render a book account of his most important flight and discovery. His flight held greater meaning than the combined journeys of men which history records of man and his most brilliant conquests.

Need it be asked why such an historic journey beyond the Earth was never adequately described? Who, including the famous admiral, was capable of describing the flight's import? Has science, as an organization, ever been known to attempt description of that which it does not comprehend? Could government officials have made plausible the actual existence and meaning of the unknown land extent beyond the North Pole point? Would the meaning even now be expressed, except for this present account?

[2] During our research, we Heathens were able to discover manifold mentions of this alleged newspaper account, but not the actual article or source.

ch. ends p. 170

An incident conveys something of the flight's import. Immediately after the flight account was heard in Washington, the office of United States Naval Intelligence conducted a wide investigation of the author of a work which had described such unknown land and the reason for its existence twenty years before it was discovered. Needless to say, the author did not need such investigative attention to attest to the authenticity of his 1927 disclosures. He need not have lived to know of the memorable flight and confirming land discovery; he would still have departed this life with knowledge that the land of his premature disclosure did in fact exist.

That 1947 discovery of land beyond the North Pole point and the interest expressed by a responsible government agency should bring into sharper focus the absurdity of conjectured "spaceship" accomplishments. There would have been no interest in the land beyond unless there was some discernment of that land's possibilities for journeying into the apparent "up" points of the universe about us. Modern discovery of inestimable land extent beyond the North Pole and South Pole points of theory attests to the complete lack of necessity for "spaceships" for modern journeys into the celestial areas of the Universe.

The idea of "spaceships" and their hoped-for accomplishment is based entirely upon an archaic theory now proved fallacious in the extreme as a result of newly discovered factual values described here. An outstanding factor for the defeat of "spaceships" accomplishment is the word of theory "gravitation." "Gravitation" is a word which has value only to the conjectures of theory; it has no relation whatever to cosmic reality. The cosmic force is magnetism, not gravitation. Yet a word of theory which is opposed to cosmic reality has been accredited as a cosmic feature in order to sustain a very factual "spaceship." But as gravitation has value only within the framework of conjectured celestial mechanics, how can it possibly be utilized as a medium for accomplishment in a world of reality?

Published accounts of hoped-for "spaceship" accomplishment fantastically hold that "spaceship" locomotion may be derived from nonexistent power elements in the stratosphere. The elements are claimed to exist so as to sustain the conjecture of "spaceship" performance. We may even grant the existence of requisite power elements. Yet it becomes incompatible with reason to grant credence to astronomical assumption of fantastic distances and other illusory astronomical features, and at the same time to hope to journey to any celestial area by "shooting up," or out, from the Earth's surface.

There is no doubt whatever concerning the ability of mechanical engineers to construct a "spaceship" that could be elementary. But what then? Whether "spaceship" travel is embraced by reality or is nothing more than pure fiction, the developments of our time negate the necessity to attempt such journeys to areas of the universe about us. The land endlessly extending beyond the Earth's assumed northern end may be considered a celestial land continuous with the Universe area called the Earth. The celestial joins with the terrestrial at the polar barriers that man erected. Though such man-made northern and southern barriers to the celestial have for many centuries proved most formidable, modern discovery shows that they possess no greater isolating value than the wire-fence barriers erected to isolate a ranch area from its neighboring ranch throughout our western United States, or than the border between two nations.

This present application to the discovered land beyond the North Pole revives the question that arose when land extent beyond the North Pole and the South Pole was first disclosed to various scientific and academic groups during lectures of 1927–30. The most popular questions of that time were "What are the connecting links composed of beyond

ch. ends p. 170

the North Pole and South Pole?" and "Is the material flexible that joins our earth with celestial areas beyond the North Pole and the South Pole?" Such questions correspond with inquiry concerning the consistency and flexibility of Atlantic Ocean and Pacific Ocean water. The oceans serve as connecting links between the eastern and western sides of this terrestrial area; they join the eastern "Old World" with the western "New World."

The questions were not inspired only by the sensational nature of the disclosure that connecting land exists between the terrestrial and the celestial. They were impelled mostly through the inflexibility of concept developed from the fallacious "isolated globe" Earth and its illusory "circling" in space. Naturally, the concept of Earth's isolation and its isolated movement through space precludes the possibility of anything but space beyond the assumed "globe" Earth's northern and southern assumed ends. Concept has to harmonize with theory, and theory has to prescribe land's end at the mathematized geographic centers, the imaginary Pole points. Such points must of mathematical and conceptional necessity designate the modern northern and southern "dropping off" points. They are equivalent to the eastern and western horizontal Earth ends considered to be "dropping off" points prior to the discovery of this "New World." That earlier concept created fear that ships sailing to the eastern and western horizon points would "fall over the Earth's edge" and be forever lost in space, whereas the superseding concept created fear of being lost in space beyond the illusory northern and southern ends of the earth. Such is the power of concept.

Review of facts discloses that the concept developed by the astronomer Ptolemy is based in the illusory, and that this vast Western so-called Hemisphere exists where space was conjectured. The course of journey from the eastern half of the terrestrial width to the western half never necessitates shooting up or down. From one side to the other side is on a direct line.

But the globe symbol fosters the illusion that one side is under the other. "Up" and "down" are always relative on a terrestrial plane. Likewise, are "up" and "down" relative in the Universe whole. Hence the undeniable facts of modern enterprise attest to the similarity of yesteryear's conceptional error with that of our time. Yesteryear's illusions are repeated, but they have been applied to different areas.

The memorable discovery of land beyond the South Pole, on December 12, 1928, and the subsequent discovery of land extending beyond the North Pole, in February 1947, confirm that the previously assumed Earth "ends" continue into celestial land areas appearing "up" or out, from terrestrial level. Because of the structure of the Universe whole, wherein the terrestrial is factually embraced as a connected area rather than as an isolated unit, no "shooting up," or out, from terrestrial level is required for immediate and unfailing journey into areas of the celestial. The northern and southern land courses into the universe about us, to Mars, Saturn, Neptune, and every other astronomically named or unnamed area of the celestial, are now clearly defined. They can be traveled over as readily in this modern era of airplane speed and simple radio communication as an ocean steamer can move "down" from one side of the misconceived "globe" Earth or "up" from the other side. The "up" and "down" perspectives have no factual value in an ocean steamer's movement or an airplane's flight from one side of the Earth to the other side. The globular concept fraudulently attires such perspective with reality.

In an attempt to view the Universe and to determine journeys to its celestial areas, the relation of the terrestrial to the celestial is likewise provided with false "up and down" isolation because *the celestial appears to be up* from the terrestrial. Hence the seeming, the apparent, the deceptive

ch. ends next p.

condition becomes endowed with realism in plans for journeys to celestial areas. Though the error of concept may be understood and excused, it can in no way be modified unless the concept is discarded. Conceptional growth is ever dependent on the nourishing irrigation of change. The following comparison is provided as a timely irrigant[3] conducive to such change.

[3] A fluid used to wash away debris, bacteria, or other unwanted materials from a wound, body cavity, or root canal.

10
A Comparison of Values

An insect is endowed with human intellect and sight. Its habitat is in the center of a waving or undulating flag, or banner. Every area of that undulating flag or banner would have to be considered "up" to the insect's observation. That condition would prevail despite the fact that every observed and unobserved area of the banner or flag is on the same level as the area where the insect dwells. Regardless of where the insect moved, from its relative "down" position in relation to other locations appearing "up," the position it had abandoned would have to appear to be "up" from every new position acquired in the insect's Universe. The flag or banner is the insect's Universe.

In like manner is terrestrial man related to all celestial areas of the Universe. "Up" is everywhere. "Up" is from every angle of observation on man's terrestrial area. It is the same for celestial man; the terrestrial which terrestrial man considers "down" or under the celestial is "up" to observers on the celestial.

Though the banner does not describe the realistic arrangement of the Universe whole, it may assist human intelligence on a terrestrial plane to determine the relation of position to other areas of the Universe whole.

Hence when we journey straight ahead from our assumed

terrestrial "ends," we will continue to be moving on the same physical level with the terrestrial area of our present knowledge. But in that movement on the same level, we will in fact be progressing into the celestial areas which, from terrestrial observation, must appear to be "up." On our arrival at celestial location five thousand and more miles beyond the assumed terrestrial ends, terrestrial sky areas of the Temperate Zones and the Torrid Zone will appear to be "up" from our celestial locations beyond either Pole point. As the terrestrial areas will telescopically appear to be "up" from the new location beyond the Pole points, the luminous sky over all celestial areas other than the area of our occupancy will likewise appear to be "up" in relation to our newly acquired celestial position beyond the Earth.

Now consider the measure of confusion to develop in the insect intelligence when its banner Universe is moved into a dark environment, and the entire Universe area receives a coating of luminous paint. The luminous coating is so applied as to leave certain areas so thinly coated that, in comparison with thickly coated areas, the luminous content cannot be detected. The luminosity will be so vague, in comparison with that of other areas, that the vague areas will not be considered the same as the more luminous areas.

The sight of the insect, being equivalent to that of human beings, will add to the confusion, by lens development of every banner Universe area into a deceptive globe or sphere. With that development, the vacant areas of the banner Universe will be multiplied.

Would not the insect intelligence be compelled to conclude that there exist innumerable globular or spherical areas constituting its banner Universe? And would not the insect concept hold that space exists between areas of the banner Universe? It would be inevitable that the insect be confronted with space, though no space exists in fact between areas of the insect's banner Universe. The insect, like its human creators, possesses visual ability which causes lens convergence. And that lens function demands that every luminous area of the banner Universe deceptively appear as a globular and therefore isolated "body."

It may be timely to repeat: When it is stated that the area would deceptively appear globular and isolated, "it is the brain that truly sees." Hence, though the detecting lens would find disk areas, the disk area detected automatically becomes a globular and isolated entity to the mind. In common parlance, "mind" is synonymous to "brain," though in reality the mind is the development of brain functioning. However, the result is the same. The lens detects the disk area in fact the lens creates the disk area. And at once the brain interprets the disk area of lens development as a globular "body."

As an additional feature to confuse insect intelligence in observation of its Universe, we would impose between the insect's sensitive optic lens and its numerous luminous banner Universe areas all the chemical elements confronting human observation of luminous celestial areas. How accurately can the insect be expected to determine realistic values of the deceptively globular and isolated luminous areas of its banner Universe when observation is influenced by the known factors influencing human observation and conclusions? Some of the influencing agents are as follows:

1. The insect's immediate blue sky would be in constant gaseous turmoil between the insect lens and all banner Universe areas. And the same influence would exist for any aiding telescope lens the insect might manufacture.
2. The luminous outer sky surface might project radiation in the stratosphere, depending upon conditions prevailing.
3. Beyond the luminous outer surface of the insect's blue sky, the constant and erratic movement of cosmic rays

ch. ends p. 179

would interfere with insect observation and influence insect determination of observed banner Universe areas.

4. Another influencing agent would be the ultraviolet rays from the Sun.

5. Other particles from the Sun would also influence observation and conclusions. Such particles, restricted to stratosphere performance, would be dual agents; they would be present in the stratosphere over the insect's immediate sky, and over the luminous sky area under observation by the insect.

6. The radiation from some observed luminous area would, under certain conditions, be reflected in the stratosphere over observed areas. That would contribute another element of confusion.

7. The continuous movement of sky gas on the observed luminous area and the variation of that movement would create all manner of illusions.

8. Variation of the brilliancy of many luminous sky areas would impose further hazard for insect determination of values.

9. And God help the insect intelligence, were it to add to common lens errors the gross deceptions which would result from telescopic magnification of banner-Universe luminosity. Thereby would be developed all the grotesque entities presented by light shading and light distortion.

The human intelligence creating the insect and its banner Universe will know that the banner Universe is finite. Therefore, it may fail to realize the insect's plight. So that creating intelligence may more fully comprehend, we need but lengthen the banner Universe so that the beginning and the end are not to be observed or determined by human intelligence or insect intelligence. Thus the original limited, or finite, banner Universe we created for the insect becomes and endless structure sweeping through infinity's darkness. It may be likened to an endless plain that is at times known to envelop one during the dream projections of sleep. And it is within the bounds of conscious projection.

Now, we are only deputies of the Supreme Intelligence. It alone may know the beginning and the end of the banner Universe. We are restricted to discernment of the insect's plight on the immediate banner Universe we know. That area and its conditions are known to our creating intelligence, but the insect is denied such knowledge. Though we may more readily determine the insect's problems on its immediate finite banner-Universe area, we cannot determine the end, which has become out of bounds for us as well as for the insect.

The particular banner-Universe area we know better than the insect corresponds to our terrestrial area of the greater Universe whole. So let us assume that we watch the insect's attempts to reach his "Heavens above," which seem to be directly overhead from insect observation. Over a period of time we view the insect's flight up from its banner-Universe location. The insect always is propelled back to an area of the banner Universe removed from the point it started from. Finally, with unprecedented speed, the insect makes a desperate effort to attain the points *apparently* overhead. And the insect does not return to any area of its banner Universe. It misses the apparent overhead points, and it misses all areas of the Universe. It goes beyond the Universe structure.

Accordingly, we create an insect replacement. With direct knowledge of the original insect's error of procedure to reach apparent "up" points, how should we advise the new insect inhabitant of our banner Universe? Should we advise it to "shoot up," or out, from its banner Universe location, thereby taking it away from the Universe structure and points thereon it desired to reach? Or should we, with

ch. ends p. 179

broader view of the banner-Universe, advise the insect to move straight ahead from either end of the banner-Universe area originally designated for insect abode?

Naturally that insect area would have length and width, the same as all other banner-Universe areas. It cannot be conceived as a mere point to accommodate the ordinary insect on any commonly known point, such as wall, floor, or ceiling. This is an extraordinary insect; it must have exceptional living area.

That insect area on the banner Universe corresponds to our terrestrial area in the greater Universe representing the Creation. So despite the fact that insect progress would be barred by apparent dark and vacant space between its permanent location and the luminous banner-Universe areas it desired to reach, should we not reasonably advise that there be no shooting up?

As it would be to the insect on its area of the banner Universe, so it is with terrestrial man on his area of the greater Universe. Remember, that the flywheel Universe described in **Chapter 2** is intended only as an illustration (**Figure 1**). The Universe is not constructed in the manner of an enormous pinwheel. Nor is any area in fact isolated from its neighboring area.

Though the flywheel does not show the realistic contour of the Universe structure, the *realistic contour* of the Universe structure *is contained in that illustration*. A hint should be sufficient for comprehension of the Universe contour in the space where it was created. If the hint does not serve, modern civilization is not entitled to know the structure of the Universe. Previous pages have disclosed that it is not so much what one sees but, rather, how it is seen. The structure of the Universe is shown in the illustration, but it is not shown by simple view of the illustration in the form depicted. "None are so blind as they who will not see." Therefore, if one would see, one should look in many ways and from many different angles.

In view of the painful knowledge of the globe *symbol's* magic power over average concept, the handy flywheel illustration was considered the most suitable means to describe how a physical journey can be made into the celestial from beyond the North Pole and South Pole mathematized ends of a supposedly isolated globe Earth. It adequately indicates the illusion of globularity of celestial and terrestrial sky areas. At the same time, it expresses Physical Continuity of the terrestrial with the celestial. The author knows that inherent in that illustration is a factual model of the Universe structure.

Figure 1, the flywheel illustration, was inspired, at least in part, by the response of earlier lecture audiences. The response disclosed that concept of our Earth's Physical Continuity with celestial areas is more readily acquired through visualization of the northern and southern terrestrial extensions as globes added to the original terrestrial "globe."

It was also disclosed then that comprehension will develop from visualizing the Universe whole as a series of connected cylindrical areas. That visualization does not have as sharp an impact on concept because it does not represent so drastic a departure from the globular. Any area of a cylinder can be drawn to globular proportion.

The most difficult problem for the average concept thirty years ago was that of supplying flat land surface to the land extensions beyond the Pole points. The problem should now be resolved, with knowledge that this nation has bases established on the land beyond. With modern discernment of values in a world of reality, one need not question the course of Rear Admiral Richard E. Byrd in February 1947. That course extended for nearly two thousand miles beyond the Earth. And if the feature was not widely heralded, there is nothing subtracted from the accomplishment. It is certain that there can no longer exist doubt concerning the physical

ch. ends next p.

reality of surface land, and mountains, and lakes, making the spaceless course of the admiral's flight beyond the Earth.

In the light of current research and modern discovery, what precisely ordered concept of organized science can be expected to challenge successfully the presently confirmed claim of thirty years ago that such indeterminable land and water course extends into the universe about us? What value can attach to yesteryear's mathematical theories of the Universe in the light of irrefutable modern discovery by accredited United States scientific research and explorative agencies? Their findings deny theory's premise. And they establish the earth as a Physical Continuity of the luminous celestial areas about the Earth.

What difference does it make if the Universe whole was created in the form of an enormous flywheel, or as an endless cylinder, or as a banner or a plane with sweeping extent beyond the bounds of mortal concepts? No mortal, as a mortal being, will ever be privileged to leave the Universe structure and thereby to view its movement, if it moves. One cannot photograph the motion of a train on which one is riding. But one can leave the train to accomplish optical observation and photographic recording of the train's movement.

Words and phrases of conjecture concerning Universe contour and movement are toys for childish quibble. The world held an abundance of the conjectural word before fact-finding instruments permitted the sensational discoveries recorded here. And one simple little discovery of infinitesimal fact is capable of dispelling countless centuries of wild and meaningless conjecture.

The most important thing for the demonstrative insect created on the banner Universe was how to reach other equally substantial areas of its Universe. It was denied access to other areas as long as it theorized upon the course presented by deceptive appearances. The insect's relation to other areas of its miniature Universe would in no way be changed if its Universe had been constructed as the illustrative flywheel Universe or as a cylindrical Universe. The same illusions would exist. And the insect would encounter the same infinite space if it developed sufficient speed to keep it from returning to other areas of its Universe. But the insect would not accomplish journey to apparent "up" points of its Universe by "shooting up."

That which applies to the insect on its miniature Universe also applies to man on his terrestrial area of the greater and realistic Universe representing the Creation. Man cannot "shoot up," or out, to celestial areas which are *apparently* up from the terrestrial.

The concepts of a connected and continuous spherical flywheel Universe and a cylindrical Universe whole may be considered of corresponding value. But both present logical Universe patterns capable of explaining experienced terrestrial conditions which inaugurated the Copernican Theory. Such conditions are the long days, the short days, and the seasons of the terrestrial year. Both concepts are stripped of the illusory, which was basic to the Copernican Theory. And they permit immediate acquisition of celestial land areas, whereas the Copernican Theory can never permit movement from the terrestrial to the celestial.

In addition to the depicted flywheel Universe whole, the undulating banner Universe constitutes another distinct concept of the Universe structure. Both are opposed to the illusory "isolated globe" concept of the Universe whole, yet they very capably explain experienced terrestrial conditions while affording the definite advantage of providing a course for immediate journey into the universe about us.

11
The Magnetic Respiration
of the Universe

Areas of the flywheel Universe shown in **Figure 2** could readily be drawn to the cylindrical. Then every connected land area of the celestial and the terrestrial could be undulating through the power of every area's magnetic energy. The undulating would be toward and away from the Sun, and the Sun would be moving in its unchanging course along the entire Universe structure. The Sun's perpetual journey along the Universe course would be that of paternal supervision for the Universe whole.

Hence the Earth's daily movement, in conjunction with similar movement of all celestial areas, would be toward and away from the Sun's path. Such movement would account for day and night. The Sun's continuous movement along the Universe course would at one season of our terrestrial year be toward the terrestrial area; at another period of the year it would be moving away from terrestrial areas. Moving toward and away from the terrestrial would be equivalent to the Sun moving in the same course but moving slower in the summer months than in the winter months. And both conditions would be equal to the Sun's remaining always on the same course but dispensing greater solar energy at one season of the year.

Either of the three conditions will adequately explain the experienced seasons and the longer and shorter days of our terrestrial year. The same conditions could produce the same results for other areas of the Universe whole. They, too, experience long days and short days, and seasons, and their periods of night and day vary.

This undulating movement of the terrestrial area and all other inseparable areas of the Universe whole may be likened to the individual's breathing, or expansion and contraction of the lungs. There are known variations in the speed, or intensity, of individual breathing under normal conditions. And there is at other times an abnormal breathing which may be drastically slower or faster, depending upon individual condition. Accordingly, there develops constantly varying speeds of breathing among *all* the Earth's individuals at all times.

Areas of the Universe would also express multiple variations in the speed of their daily undulating movement toward and away from the Sun's course in space. And the variations in movement of areas of the whole would be consistent with continuous unity of the Universe whole.

Normal breathing of individuals has a range of from fourteen to twenty-four cycles each minute, whereas under extraordinary conditions, particularly in cardiac and respiratory diseases, the number of breaths, or respirations, may be increased to fifty or decreased to eight. Therefore, it may be discerned that there exists constant variation of the speed among all terrestrial individual moving toward the same place, continued living. Each individual of the terrestrial may be considered a distinct area of humanity, and each individual attains the daily and yearly goal with varying speed of movement.

All land areas of the Universe whole may move with

ch. ends p. 206

different speeds at different times and each remain an inseparable part of the connected Universe. The terrestrial area's daily lung expansion, or partial undulating movement toward the Sun's course, could be of approximately twelve hours' duration. The terrestrial daily lung expansion could begin at about twelve o'clock midnight and attain maximum daily expansion at about twelve o'clock noon. That maximum expansion would bring the terrestrial to a space location where the Sun would *apparently* be directly overhead. Then for the next twelve hours there would be Earth breath contraction. It would complete the daily undulation, and it would return the Earth to its lowest point in space and most remote from the Sun's course. That would be the point at approximately twelve o'clock midnight.

About the middle of the terrestrial daily lung contraction, or movement from the highest twelve-o'clock-noon point in space, darkness would begin to envelop terrestrial areas. The approach of darkness would be experienced at some terrestrial points as early as 4 p.m. and at other points as late as 8 p.m. And it would result from the increased distance of such terrestrial points from the Sun's course in space.

Then the next day's terrestrial breath expansion would bring to some terrestrial points at 6 a.m. the so-called dawn. The light of dawn would increase until the breath expansion reached its peak at about twelve o'clock noon. The peak of expansion would bring the terrestrial to its highest point in space, where it would receive most of solar energy. Daylight would prevail for part of the period of terrestrial lung contraction as the terrestrial moved away from the highest Sun point.

It may be seen that as each area of the terrestrial reached its high point in space it would recede with the daily contraction. But another point of the terrestrial whole would take the high-point position vacated. Hence when the Boston, Massachusetts, point of the terrestrial was far removed from the high point of its area's daily breath expansion, Hong Kong, China, and other terrestrial areas would be approaching the high point. Some areas of the terrestrial would experience noon while others were experiencing total darkness. The same condition would apply for all celestial areas of the Universe whole. They, too, would be expanding and contracting in common with all terrestrial areas.

So, as it is with the individual human body's respiratory variations, the daily expansion and contraction of all areas of the terrestrial and the celestial would correspond. Naturally, the daily respiration of Universe areas would be much longer than a human body's respiration.

Though ever bearing the same relation as inseparable parts of the universal daily undulation in space, some terrestrial and celestial areas would move toward the Sun's course in space at greater speeds than others. That condition would develop variations in time of arrival at the high and low space points representing complete expansion and complete contraction. The difference in speed of movement would in turn develop different hours and minutes for the various terrestrial and celestial areas to experience the Sun as being *apparently* directly overhead. It would likewise develop for terrestrial and celestial areas of the universe whole variations in midnight. The arrival of terrestrial areas at the lowest space point most remote from the Sun's course would not occur at the same time.

Hence it can be understood that twelve o'clock noon would not apply to all terrestrial areas. And that would hold regardless of what movement the Earth and the Sun prescribed. Many terrestrial and celestial areas would factually experience noon at different hours and different minutes of the hours. The theory of globular isolation makes allowance for such difference in time, but twelve o'clock noon is accepted throughout the terrestrial as a matter of convenience.

ch. ends p. 206

The following word illustration of a comparative movement to be observed at terrestrial level seems pertinent and
may assist visualization of the daily universal undulation.
One can readily visualize a frail craft as it participates in the
rolling water motion of a calm lake or river. Visualization of
the same craft obliged to participate in the violent wave and
upheaving motion of a turbulent body of water will materially assist mental comparison of values. It can be discerned
that in the case of turbulent water the bow[1] of the frail craft
may be almost upright in space while the stern[2] could be on
the surface of the water. Thus the craft would be seemingly
standing on end. And every other area of the craft whole,
from bow to stern, would occupy a *different position in space
while retaining unity with the craft.*

The mental portrait of terrestrial and celestial land areas
making a unified daily undulation on the broader ocean of
infinite space may be enlarged as follows. Add to that single
craft a hundred or a thousand similar craft. Have the bow of
one scraping the stern of a connected craft along the entire
length of undulating craft comprising the whole. Each craft,
and every part of each craft, would reach its necessary high
point, or position, in the relative space where the undulation
prescribed. The highest point to be attained in space away
from the water's surface would not and need not be the same
for all parts of all craft comprising the undulating whole.

Each craft and its various parts would *in time* return to
a momentary position of even keel, or near even keel, on
the surface of the water. And any change of speed for the
undulation would affect the time spent by various parts of
the undulation at the low-water surface point and the high
space point. The highest point to be attained in space away
from the water's surface would correspond to the highest
point to be attained by certain terrestrial and celestial areas

[1] Front.
[2] Rear.

in their daily undulation toward and away from the infinite space path representing the Sun's course. And the lowest even keel or near even keel, position to be attained by all boat parts would be symbolic of the terrestrial and celestial land areas' lowest point of daily undulation toward and away from the Sun's course.

Some parts of the combined terrestrial and celestial, or areas of the Universe whole, would, like the undulating boats, reach the high space point simultaneously. But never could all parts of the unified terrestrial and celestial daily undulation attain the high point at the same time. The same holds true for the completion of the daily undulation which brings every area of the Universe whole to the lowest point in space away from the Sun's course. That point would be midnight, but under no circumstances could it be twelve o'clock midnight for all areas of the combined terrestrial and celestial at the same time.

Accordingly, all up movement to the highest, or Sun's course, point in space infinite would represent the daily course from midnight to noon for terrestrial and celestial land areas. The second phase of the daily undulation would be away from the high Sun's-course point in space toward the lowest point in space. That would be movement from the high noon point to the low midnight point. Time would have no bearing on the situation. Though the time of arrival at high point and low point would vary for areas of the undulation, the arrival at high point would be noon for each area, and arrival at low space point would be midnight for that area.

The foregoing demonstrates how day and night could be experienced without any necessity for isolating the Earth and other realistic land areas of the Universe. The Universe can survive as a unit, and every land and sky area of the Universe can continue to be connected. Yet every area of the

ch. ends p. 206

Universe whole can prescribe daily movement toward and away from the Sun.

In a consideration of the seasons, it is shown that the Sun in its yearly course would be directly over some terrestrial and celestial areas at certain periods when the Universe whole was prescribing its undulation toward the Sun's course in space. The undulating approach of various Universe areas to the Sun's course would not imply that the Sun was in fact overhead. For the majority of Universe areas the Sun would be anywhere but overhead for the greater part of the year, regardless of the Sun's *apparent* overhead position.

Direct relation to the Sun and direct relation to the Sun's course are quite different conditions. The former provides direct perpendicular benefits from the Sun when it is the least distance away from a particular Universe area, whereas the latter condition would permit only of the Sun's benefits at any angle. And the increase of angle would increase the Sun's distance from a particular area.

Therefore, the day's length and the seasonal change of areas would be influenced not only during the brief yearly weeks when an area had direct perpendicular relation to the Sun in its course. The change would also be felt for a period while the Sun in its course *was approaching* direct perpendicular relation to any area as well as when the Sun was moving on its course *away from* a particular area.

As the Sun moved in its course there would develop for other terrestrial and celestial areas the same seasonal change. It would be experienced as the Sun approached to perpendicular relation with such an area, and the change would be most marked when the Sun had reached direct perpendicular relation. Then, as the Sun continued on its course away from perpendicular relation with the particular area, there would develop another seasonal change for the particular area. Thus some terrestrial and celestial areas would be entering their summer season while numerous other terrestrial and celestial areas would be entering their winter season. Some areas could be simultaneously experiencing the longest day of summer while other areas, receiving Sun benefits at the angle when the Sun was most remote from such areas, would be experiencing the shortest day of winter. There would thereby result variations in the exact time of direct Sun for the different Universe areas along the Sun's course of inconceivable extent.

Such could be the development between the Sun and all areas of the Universe whole, even while every area of the whole could be *seemingly bearing the same daily relation to the Sun.* However, that seeming condition would develop from each Universe area's daily movement *toward and away from the Sun's course.* Though a particular area might seem to be approaching the Sun, the Sun could be at its most remote Sun-course point from the area. The appearance of direct Sun could prevail at such a time and place, but the benefits of direct Sun relation would be absent.

The terrestrial equatorial area, and the corresponding celestial equatorial area, or Torrid Zones would result from the fact that such areas would reach the highest point in space on the universal undulation making for the day and night of all areas. But the Torrid Zone areas would never reach the lowest point in space, most remote from the Sun's course, to which Temperate Zone areas would be obliged to move. Like the undulating boats of the illustration, the Torrid Zones of the Universe would reach only the point of obliged through the function of the universal undulation to reach absolute even keel. Even keel for areas of the boat the water's surface. Even keel for land areas of the Universe making daily undulation would mean the lowest point of the undulating in space.

The daily participation by Torrid Zone areas in the universal undulation toward and away from the Sun's course would be sufficient to ensure day and night change for such

ch. ends p. 206

areas. But because of the added advantage of *their location* on the universal undulation, their daily movement, or dip, away from the Sun course and toward the lowest midnight point in space need not be as sharp as that of other areas. And their speed at such time and place could be increased so that they would get away from the low space faster than Temperate Zone areas of the terrestrial and the celestial do. That feature would provide a time advantage in approaching their daily high point toward the Sun's course. Hence for all Torrid Zones throughout the Universe there would be shorter nights and greater warmth. And there would not be the marked seasonal changes of Temperate Zone areas.

On the other hand, the Frigid Zones, or polar areas, of the Universe would hold such placement on the universal undulation that they would have to reach the lowest-possible space point. And the undulating movement of their particular part of the universal undulation would be barely perceptible when compared with the movement of other areas. Hence for half of the year their movement of ascent in space to the highest point approaching the Sun's course would be negligible. And it would result in the six months of darkness, and near-darkness, characteristic of Frigid Zones. During the other six months of daylight, or approximation thereto, the same frigid areas, terrestrial and celestial, would hold a relatively stable position toward the highest universal undulation point in space. The position of frigid areas during that period of universal undulation would provide proximity to the Sun's course, permitting sunlight to prevail. However, during the period of high-space-point occupancy, the frigid areas having six months' daylight would not experience a direct perpendicular relation to the Sun in its yearly course. No area of the Universe can experience that period of direct Sun relation. But the continued six months' propinquity[3] to

[3] In this context: relationship; kinship.

the Sun's course would be sufficient to provide the condition of enduring daylight.

Therefore, though the frigid areas of the Universe would have sufficient summer angle relation to the Sun for a measure of daylight beyond that of other areas, they would be deprived of direct overhead relation to the Sun during that period. Hence they would not be provided the measure of heat lavished upon tropical and temperate areas during a part of that same period. In other words, as the frigid areas held their highest undulation point, or proximity to the Sun's course, it would not represent the high space point of Temperate or Torrid Zone areas. It would permit reception of sufficient Sun force to ensure continuing light, but the angle of that reception would prohibit the intensity of heat received by Temperate and tropical areas during part of the same period, when they were at their highest point of the undulation.

There would be other conditions influencing seasonal changes of the year for terrestrial and celestial areas participating in the perpetual universal undulation toward and away from the Sun's course in space infinite. There may well exist the very definite influence that would result from lack of consistency in the Sun's dispensation of energy which produces light and heat, or at least substantially contributes thereto, over terrestrial and celestial land areas. It could be that the Sun's dispensation of energy varies from time to time. At times, some areas of the undulating Universe whole would be receiving less of solar energy than at other times. Such a condition could develop from the fact that, as certain areas reached their Sun's-course point of summer, the Sun would be emitting less energy than it did when other areas arrived at a corresponding position in space. That factor would offset the benefits such areas would normally receive as a result of their direct relation to the Sun in its course.

A comparable condition could influence the winter period

of various terrestrial and celestial areas. They could be benefited by the Sun's increase of energy dispensation, and there would be modification of the winter cold of such areas.

There is no criterion that dispensation of solar energy does not vary in quantity and/or quality. But there is every indication to sustain the premise of periodic change in the Sun's dispensing of energy. Hence the location of connected Universe areas on the universal undulation, and their angle of relation to the Sun, would influence climatic conditions, seasonal change, and the length of days. The speed of movement in attaining and holding high and low space points, nearest to and most remote from the Sun, would likewise contribute to seasonal change and the length of terrestrial and celestial days. And the periodic difference in the measure of dispensed solar energy would also merit consideration as an influencing agent.

Another complicating possibility is that the Sun, while making its yearly rounds of the Universe along its course, performs a secondary movement away from and return toward the constructed Universe whole. That would make for periodic increase of distance from terrestrial and celestial areas to the Sun's course. Hence some Universe areas could be expected to benefit and others to lose benefits by the secondary Sun movement changing the Sun's course. It would depend on their location in the Universe whole.

Accordingly, to consider a secondary movement by the Sun, the conditions to develop from a difference in the Sun's dispensation of energy could be expected to develop even though the energy remained constant at all times and for all areas of the Universe whole. That secondary movement would be the equivalent of periodic modification and intensification of energy dispensation.

Further, the Sun may veer from its course *in conjunction with* periodic modification or intensification of the solar energy dispensed. There is no criterion within the extensive domain of astrophysics and its assumptive mathematical values to deny such possibility. Infinite mathematics may reign supreme in the Universe of the mathematician. And they may dictate the functions of such Universe. But the Universe of their application has been proved alien to reality by realistic modern performance. Astrophysics has no formula for the directional activity of cosmic rays within our immediate stratosphere area of infinite space. And since that stratosphere area is only the distance of a few minutes' journey over the Earth's surface, there certainly cannot be real determination of energy dispensed by the Sun at its assumed distance. And if a gauge of the solar energy dispensed was to be had, it could have application only to the time of measuring the energy dispensed; it could not gauge the energy dispensed ever over a twelve-month period. And the gauge could apply only to the immediate area where measurement was made. By no stretch of the imagination could it be considered to apply to all the areas of the Universal whole.

In view of archaic theory's assumed movements of an illusory globe Earth, there is nothing sensational in the possibility here projected that the Sun may perform a secondary movement. To sustain a postulate which isolates the Earth and disrupts the realistic Universe, the Earth is considered to make a primary daily movement on its imaginary axis at the rate of one thousand miles an hour. And it is assumed to make a secondary movement in its yearly course toward the Sun at the rate of six thousand miles an hour.

Observe the flywheel Universe in **Figure 2**. It is stripped of the illusory lens-produced curves shown for the inner and outer sky areas of its companion **Figure 1**. It conveys how the free Sun could veer from a direct space path during its yearly course over the constructed Universe whole. And that periodic departure from course could take it any number of miles away from the created Universe. There is no way

ch. ends p. 206

of illustrating where the temporary Sun path would be, but the secondary movement away from the Universe would be in a way realized by drawing a line from the illustration's stratosphere center toward one looking at the illustration. There would be no purpose in drawing the line from the stratosphere center toward either side of the Universe illustration.

Therefore, with proper application to the physically connected and continuous Universe of **Figure 2**, in which the globular deceptions of **Figure 1** have been eliminated, one will be able to visualize every land surface area of the Universe undulating toward and away from the Sun's course in space. That Sun course may be considered to extend through the center of the illustration. From the point where the Sun is shown at the top of the "flywheel" it would move through the dark stratosphere area of the illustration. It would travel the entire length, and it would then return along that length. Regardless of what the Sun's precise position may be, every undulating area of the Universe whole would retain its relation to and physical continuity with the Universe whole and to the universal undulation toward and away from the Sun's course. The results would be the same if the Sun's placement were in the center of the dark stratosphere area of the illustration, from which point it would complete a yearly circling of the illustrated Universe circumference. Regardless of precise Sun course, the daily undulation of all Universe areas would cause it deceptively to appear that every area was circling around the Sun as an isolated unit of the Universe whole. The undulation movement of Universe areas would cause the illusion of circling around the Sun to persist regardless of what the Sun's location in space might be.

Apt parallel to that experienced illusion of "circling around the Sun" is found in a local condition. One can ride a roller coaster moving with great speed up and down, or toward and way from, a huge arclight in proximity to the undulations of the coaster. Each speedy approach toward the light, and departure from the light, must create the illusion of movement around the light. Such example is elementary, but it staggers concept to grasp the greater speed of the universal undulation toward and away from the Sun arclight with a magnitude beyond concept.

In terminating this word portrait of the connected and continuous Universe and its motion, it seems fitting to relate that the Sun shown in the Illustration will be red when observed against the perpetually dark background of the space existing beyond sky areas of the Universe. When one observes the Sun from within stratosphere darkness, it has none of the luminous sunlight quality to be observed from land areas: the sun is just a red disk when viewed from beyond the blue sky. The illumination develops from mixture of cosmic rays with chemical elements of the sky enveloping land areas throughout the constructed Universe whole. The result of such mixture produces sunlight and heat on all land under the universal sky.

And it is that cosmic-ray contact with gaseous sky elements that results in the luminosity of every outer sky surface area to be observed against the dark stratosphere. The same stratosphere darkness prevails over celestial sky areas as is known to prevail over terrestrial sky areas. And unless that darkness did prevail over sky areas everywhere, there would be no art of astronomy. Only the darkness permits detection of the sky-light.

We now proceed from the flywheel illustration of the Universe and its motion to the original illustration of 1928. Though the first is last in descriptive analysis, there is nevertheless a logical pattern. Presentation of the original illustration permits observation of only a segment of the entire Universe embraced by the flywheel illustration. However, it may serve to demonstrate the transcendent values in

ch. ends p. 206

land areas discovered, in opposition to centuries of scientific deduction denying the land's existence, beyond the North Pole and the South Pole points of our Earth.

To accomplish the illustration, we must first "drop back" into space both upper angles of the flywheel at the Sun position in **Figure 2**. Both angles will remain attached to the unbroken area of the flywheel circumference, but they will drop back in to space enough to permit both to project out of sight beyond the Sun's location. The remainder of the flywheel circumference area will then extend in space a streaming banner Universe on the horizontal. The Sun will then be situated over the horizontal Universe, and the Sun's course in space will be over the Universe.

Now the horizontal two-sided banner, or plain, Universe will begin a series of arching at the Sun point and the arching will continue along the entire length of the illustrated Universe area that can be held on the page. More of the Universe beyond both edges of the page will do the same, but that area cannot be seen. The series of arching up and down, toward and away from the Sun's course above the Universe, will prescribe an undulation of the Universe areas.

Every area of the banner Universe presented could readily be cylindrical. That contour would in no way interfere with Physical Continuity of the whole. Moreover, the developments in a world of reality will be the same if the illustrated Universe extends beyond the Sun, and the Sun's course is above the undulating Universe whole, or if the Sun is moved with the unseen area of the Universe which comprised the upper right angle of the flywheel. The Sun would then be at the head of the Universe undulation. It would act as leader or guide for the entire Universe structure. Then the Sun would not prescribe its yearly course along the Universe structure as described in the flywheel illustration; its course would become the course of the force it dispensed, and that magnetic force would be transmitted along the entire Universe structure. Then every sky area of the Universe would absorb whatever portion of that perpetually dispensed magnetic force it required. As previously explained, some areas would take less because their condition required less. Other areas would absorb more because their condition demanded more. Hence the inconceivable length of the Universe whole is enveloped in perpetual darkness over, or above, the continuous luminous outer sky which extends with the Universe land structure. And along the infinite course of the Universe, a magnetic force inherent in the structure serves to maintain it on the original construction plane, or level, in space infinite. That realistic magnetic force, engendered within the land structure, may be likened in its eternal function to the human body's actuating spirit. It receives constant replenishment from the Sun's dispensation of energy, which is first received in the sky over all land areas of the Universe.

That magnetic force dispensed by the Sun serves a very definite purpose in the outer sky areas where it is received. From the sky it penetrates into the depths of the land, terrestrial and celestial. But, again like the human spirit, its function is never completed. If the Universe makes any movement whatever, it is that inherent magnetic force which actuates the motion. And if the motion is that of undulation, it is the magnetic spirit of all land areas of the Universe which actuates the undulation.

That magnetic force of the Universe is beyond the bounds of theory and abstruse mathematics. Its most formidable application serves to keep alive in all realistic matter the natural creative endowment or, if one prefers, the spark of Divinity. So the shaping of a pebble on the shore, a pearl in the oyster shell, and the perfecting of a diamond, a ruby, and an emerald, or the development of a single drop of oil in the bowels of the land are no less expressions of creative ingenuity's magnetic force than the inner blue and outer luminosity of the sky which depends on that force. The up

ch. ends p. 206

rearing of a mountain at one time and place, or the obliteration of an island at another time and place, attests to the universal magnetic influence from the crater of the Sun. If all known philosophy had been rendered eternally mute at its inception, the magnificent truths of creative reality would have been self-evident as a result of the ceaseless function of magnetic force throughout the Universe.

All that was described of the magnetic function of the Universe depicted by the flywheel illustration has equal application to the presently described Universe extending as an endless plain through infinite space. The undulation of flywheel circumference areas toward and away from a central Sun would be equivalent to an undulation by areas of the horizontal plain Universe toward and away from a Sun course above the Universe and its movement. The horizontal-plain Universe is comparable to the insect's banner Universe extending on the horizontal and waving or undulating in space. And the conditions developing from both universe patterns, flywheel and horizontal plain undulating toward a Sun center and toward a Sun course, would apply to a third Universe pattern where the undulation would only *seem to be* toward and away from a Sun leader on the same level as the Universe structure.

The horizontal-plain Universe, like the Earth's realistic plains and deserts, possesses length and width. But as the length is infinite, the ends transcend conceptional capacity. Hence they cannot be subjected to physical view. However, the width of every Universe area may be established in the manner that width of this terrestrial area of the Universe whole is acquired. But the width cannot be established until after we arrive at the particular Universe areas. That consideration would have to apply regardless of the shape of the realistic Universe whole.

There is more to be said concerning width of unknown Universe areas. It will provide the answer to the contour of the Universe whole, but it is very doubtful that the answer will be seen.

It is absurd to attempt calculation of unknown celestial areas of the Universe with application of astronomical gauges. However, and without thanks to astronomy, every unknown area of the celestial universe about us is as accurately charted in width as every area of the known terrestrial. Thus, the answer to the realistic Universe contour, previously pointed to by the flywheel illustration, is again pointed to by the foregoing assertion that the celestial width pattern is shown by terrestrial width determinations.

As we return to further description of the illustrative universe, it should be borne in mind that nothing has been said about seeing the width of unknown celestial areas of the Universe whole. We will never see the width until we arrive at the particular celestial areas. But we may *know the width from a pattern to which we have access.*

In the case of the illustrative flywheel Universe, every angle thereof participated in the universal undulation toward and away from the Sun's course in space, or toward and away from the center of the dark stratosphere area of the illustration. All corresponding areas of the horizontal plain Universe would prescribe the same movement up and down, or toward and away from the Sun's course, which would be above the Universe structure. It may be observed that in both cases the Sun's relation to all areas of the Universe would remain the same. The visualization from flywheel to horizontal-plain arrangement of the Universe whole in space would in no way alter the Sun's course in space with relation to the Universe it served.

Let's check the situation. In the flywheel Universe the Sun's course would be from its depicted location through the center of the dark stratosphere area. When the flywheel outline is terminated and the circumference stretched out to a horizontal line which extends beyond both ends of the

ch. ends p. 206

page holding the illustration, the Sun's course becomes a course above the horizontal-plain Universe. No matter what words are used to explain the situation, the undeniable fact remains that the Sun's course in space is unchanged. In both cases, the Sun is *above* the Universe structure. We changed the contour of the Universe, but we did nothing to the Sun and the Sun's course.

Though the Universe contour may be known, it must ever remain beyond human sight. The realistic pattern of the created Universe could not even be seen by an observer beyond the Universe, wherever that may be. We who inhabit the terrestrial area of the Universe, and are privileged to theorize and conjecture upon the Universe contour in space, are, after all, a part of that Universe. The patterns we apply to the Universe are but timely stopgaps[4] to explain conditions and events, both factual and seeming. And the patterns imposed by our theorizing and conjecturing must be remote from creative reality.

In both illustrations of Universe contour and movement, each terrestrial and celestial area undulated out in space from the allotted position in the space where it had been created. In so doing, all were ascending toward the Sun's course then, having reached the peak of each area's daily expansion, they would return through contraction to their original positions in the created Universe whole. In such manner they caused to develop the physical conditions experienced, particularly long days, short days, and seasons, as well as the manifest conditions of day and night. And such conditions experienced at terrestrial level have to be experienced at celestial level.

However, a reasonable explanation of experienced conditions did not demand severance of one area of the Universe from its neighboring area. Nor did the explanation

---

[4] Temporary but not long-lasting solutions.

of conditions necessitate acceptance of the illusion that every area of the Universe is a globular area. And it did not require that every celestial area and the terrestrial whole be assumed to be isolated in space and hurtling in a mathematical orbit, at various fantastic speeds for the different areas, in a yearly course toward and away from the Sun.

In the light of modern discovery, the concept of globular and isolated Universe areas is discredited, and the discoveries preclude any possibility that areas of the Universe whole are "circling or ellipsing in space." Hence the undulating movement of the Universe as a connected whole presents a much more reasonable expression of creative ingenuity. And it fits into the pattern of modern discoveries. If we, as insects of the created realistic Universe, demand that it moves, let us assume a reasonable movement which affords opportunity to visit other areas of the Universe, after having conjectured how to achieve the visit for centuries beyond estimate.

As previously related, there exists not a single creative manifestation of energy at work where "circling or ellipsing" actually takes place. Though there are examples without number where such "circling" or "ellipsing" *seem* to be performed as a result of lens function and the ensuing deceptions. This consideration is not to be confused with the mechanics of man, in which a profusion of wheels and globes perform their definite function of circling, or revolving. There is no mistaking their movement. It would not be possible for them to move otherwise. But they are far removed from celestial mechanics.

The revolving man-made mechanics, expressive of man's mechanical ability, confirms all that has been related concerning the origin of man's globular illusions of the celestial. For it was the circular structure of the human lens which inspired man's construction of corresponding circular instruments. But the instruments were formed by man to require a circling movement, and no other movement. And

ch. ends p. 206

it was the structural form of the optic lens which demanded that man view every area of the Universe as globular and, therefore, isolated. Hence the Universe whole had to appear deceptively to be comprised of many millions of isolated areas.

Recent discovery confirms that the terrestrial area of the Universe whole did not escape the disease of lenses. It, too, appears as many millions of isolated globular "bodies" adrift in space. God did not fashion it in such manner. Man was incompetent to fashion it in any manner. But the lens did fashion it in *the image and likeness of the lens.*

Realistic creative expressions of energy conform to a waiving and bending motion. And a series of waves would present an undulation. But unfortunately for human progress, the waiving and undulating motion presents the illusion of circling when viewed at sufficient distance under certain conditions.

There are light waves, heat waves, sound waves, color waves, heart waves, brain waves, and others. They are, each and every one, realistic manifestations which can be recorded. Some can be seen. Others can only be detected by extremely sensitive instruments. Carried to the ultimate, there are spirit waves which, at least at times, are discernible. They can be weighed and recorded. And they can, under appropriate conditions, be *seen in transit.*

This has to do with pure energy, and its factual expressions in a world of reality. And if one might have conjured the rainbow at the mention of color waves, it should at once be eliminated from the category of pure energy. The rainbow formation is *created by the lens expressly for the lens that is observing it.* The rainbow, or any tangent of a rainbow, has parallel with the so-called "curvatures of the Earth." And as this work describes, the Earth's "curvature" must exist *for the lens* because the *lens created the curvature in its own image and likeness.*

The inclusion of spirit waves in the reference to energy manifestations seems to require some explanation. There exists an eternity of difference in the meaning of the word "seeing" as it relates to a form of self-hypnosis and as it relates to visual detection of a spirit in transit as it departs from the human body. Self-hypnosis represents the customary "seeing" of spirits, it is a mental projection rather than a visual detection. And the word "transit" should be qualified to have application only to the brief interval when the human spirit *departs the body* — that time just preceding the cessation of all bodily functions which make life as we know life. In fact, it is that spirit departure from the body which brings about cessation of living functions.

There is a facetious saying which aptly describes the development of what we term "death": "He gave up the ghost." In "giving up the ghost," the spirit departed. In this instance "ghost" is synonymous with "spirit."

However, were one to recall the spirit they have "seen," here, there, and everywhere and under all manner of conditions, the reasonable conclusion must be that such "seeing" was a conscious, or unconscious, projection of the mental image retained of a departed person's mortal body. The imagery would be of the body once living. It would not be of the spirit living in that body before the body died and the spirit departed. And the image could be of mother, father, sister, brother, wife, lover, or anyone who was known before he died. Such so-called spirits are "seen" somewhat on the order of the astronomer's "seeing" rounded bodies circling or ellipsing in space." Such "spirits," with bodies supplied by the living mind, are often seen under condition of emotional strain.

Their presence is ordained only by the mind of a living person. It is capable of projecting the body-spirit, which is not a spirit, almost anywhere. As it is mortal mind that wills the "seeing," that which is seen must be a duplicate of

ch. ends p. 206

the body image that mind retains of a former living person whose spirit has departed.

The spirit of that previously living and known body is no doubt a resident of the unknown spirit domain. And the spirit, because it is a spirit, is without physical characteristics identifying the body in which it formerly abided. The spirit cannot be a spirit and retain mortal features. Nor can the spirit have mortal mind, which was developed to serve the body's needs. The mind remains with the body. It, with the body, was ordained by the spirit which actuated the cell to build the body.

Hence we need not discuss the numerous spirits "seen" fully attired in the clothing which covered the body where the departed spirit was contained. This does not deny the evidence of spiritual attunement with a departed spirit. That is a very different matter. Under such a condition, the spirit of a living body does in fact attune to a departed spirit. Then the living body strongly feels the presence of the departed spirit. And as the brain of the living person receives the vibration transmitted by the spirit, the mind is actuated to project the body, features, and attire of *that which the departed spirit represented.* Then, much faster than the F.B.I. could function, the mind of the living person exhibits everything the living person once knew about the former living person which the spirit vibration represents.

Thus, though the departed spirit is in fact strongly felt through the spirit contained in a living body, it is the living body's mind that automatically revives from mind's storehouse of photographs a portrait of the former *body that contained the spirit manifested.* That is the only portrait the living mind holds. It cannot contain a picture of anything other than the body it once knew as a body. It has no picture of such body as a spirit.

Therefore, the living physical entity, you and I and a hundred billion others, may reasonably feel spirit presence without seeing the spirit. But how could one hope to see a spirit in body form, particularly if that body was draped in the clothing of mortal existence and expect that it could be the spirit? That kind of "seeing" a spirit expresses a form of self-hypnosis, whereas positive seeing of the luminous flash of spirit departing the body, just preceding death of the body, represents a visual function like seeing the Sun, the light, the darkness, and a million and one things and conditions in a world of reality.

The spirit is as real as the body. Without it there could be no body. It can be seen, as spirit, in its departure from the body. It has been weighed as it departed the body.[5] But it is never to be seen as a physical body. Nor is the spirit to be seen with features, and certainly not with clothing. Only the body needs features and clothing.

To progress to what might be considered a more physical realm of energy, where it is manifested in and by land and water mass, there is experienced the regular waving (waves) of oceans, rivers, and lakes. And there is also experienced the irregular expressions of tidal wave undulations. The experienced Earth tremors are expressions of underground waves of energy. They reach the peak of expression in violent undulating earthquakes and volcanic eruptions.

On the Earth's surface it is found that gases and smoke clouds billow and roll. But they do not circle. However, the billowing and rolling can deceptively appear to be a circling motion.

---

[5] A reference to what has come to be known as the 21 Grams Experiment, a study published in 1907 by Duncan MacDougall (c. 1866–1920), a physician from Haverhill, Massachusetts, who hypothesized that souls have physical weight, and attempted to measure the mass lost by a human when the soul departed the body. His experiment was carried out on six patients at the moment of their death, wherein one of the subjects lost three-quarters of an ounce (21.3 grams). MacDougall himself stated the experiment would have to be repeated many times before any conclusion could be obtained. Regardless, his experiment is widely considered flawed and unscientific due to the small sample size, the methods used, as well as the fact only one of the six subjects met the hypothesis.

ch. ends p. 206

Lightning bends, and chains, and zigzags in its course, but it does not circle. And all expressions from man's harnessing and utilizing electricity attest that the motion of electricity is opposed to circling. Where the electric current is seen as light, it vibrates to and from in an undulating manner on the filament which carries it. And the motion is anything but that of circling, even though the current is captive within a globular area, a light globe.

Wherever a true circling or ellipsing motion is prescribed, it is due to and is an attribute of man-made mechanics. And where mechanics are not man-made, as in the universe about us, man's concept imposes upon non-globular creative reality a false globular outline. There is no disputing that globes and spheres, and globular and spherical items, exist by the millions. But they exist only on the Earth's surface where man created them. And there are numerous man-made products that do prescribe a circling motion. Likewise, are there many man-made objects which, when properly arranged and provided the proper speed, will deceptively appear to be globular areas as a result of the circling motion they prescribe. Yet, when the motion ceases, it will be found that the areas are anything but circular or globular in outline.

There is available extensive knowledge concerning lens capriciousness, and the illusions known to develop from motion directly at hand on the Earth's surface. Hence it is most singular that modern man persists in endowing with reality the unrealistic globular celestial areas. And, in granting that the areas are globular, man must decree that they are isolated. Then, with false globularity and isolation in control of mind, movement detected at celestial level must be circling or ellipsing.

It is a most extraordinary development that man, after centuries of conjecture concerning the course to Mars and to all other areas of the Universe, fears to pursue the course now so clearly defined. In the initial discovery of a land course into the celestial, the existence of land beyond the South Pole was established on December 12, 1928. But the course was not then penetrated. In February 1947, the northern pathway into the so-called "Heavens above" was discovered beyond the North Pole. And a meager length of its inestimable extent was penetrated by a U.S. Naval task force under command of Rear Admiral Richard Evelyn Byrd. However, obsolete theory and the misconception it fostered for twenty-eight years restricted depth penetration of the southern course. It was not until January 13, 1956, that any real progress was made; when a U.S. Naval air unit accomplished a flight of 2,300 miles beyond the South Pole point of theory. But such extent is meaningless when it is known that journey beyond may be continued for hundreds of thousands of miles.

Nearly ten years have elapsed without notable explorative purpose over the northern course extending beyond the North Pole point. Of course, it is possible that penetration has been made beyond the 1,700-mile point reached in February 1947, but the accomplishment has been kept secret.

Could it be that terrestrial man's reticence to continue over northern and southern land areas leading into the celestial is due to the fixation of the overworked "shooting up" conjecture? In a distant yesteryear of fifty years ago, this then little boy seriously asked how far is the sky. Since then there has abided the popular thought and discussion of "shooting up" in a rocket to reach Mars, and to reach other areas of the universe about us. It would seem that at long last a more reasonable and fruitful manner of procedure might be contemplated, particularly after the modern discovery of direct land routes leading "up" from beyond the South Pole and the North Pole. Progress straight ahead from beyond the Pole points will never require "shooting up," or out, from terrestrial level to reach celestial areas.

In view of the current trend toward destruction of terrestrial man and his civilization, there is imposed the unpleasant

ch. ends next p.

thought: What a pity it would be if man were to destroy his kingdom on Earth before adequate preparation had been made for sanctuary on adjoining celestial territory. In the unwelcome persistence of such a thought, there is revived the name of a famous predecessor who dwelt in France. He was known as Jules Verne,[6] and he predicted that the Earth would be destroyed by an implement of war which would burst like a boiler. He also observed that the Americans were good boilermakers.[7] We are good boilermakers. And the instruments of destruction corresponding to a boiler are the fearful atom, hydrogen, and cobalt bombs. Can it be that, as man of this terrestrial civilization stands on the threshold of celestial land areas, and when the centuries' Dream of Dreams is about to be realized, wholesale destruction will cancel the Dream's fulfillment?

---

[6] Jules Verne (1828–1905) was a French novelist, poet, and playwright best known for his adventure novels, including *Journey to the Center of the Earth* (1864) and *Twenty Thousand Leagues Under the Seas* (1870).

[7] A reference to Chapter 16 of Verne's 1863 novel *Five Weeks in a Balloon*, wherein the following exchange occurs between to characters:

"...I have always fancied that the end of the earth will be when some enormous boiler, heated to three thousand millions of atmospheric pressure, shall explode and blow up our Globe!"

"And I add that the Americans," said Joe, "will not have been the last to work at the machine!"

"In fact," assented the doctor, "they are great boilermakers!"

12
The Master Builder
Luminous Skyprints

All are architects of fate,
Working in these walls of time:
Some with massive deeds and great;
Some with lesser rhyme.
—Longfellow,[1] *"The Builders"*[2]

Along the transcendent corridors of creative reality, architects of fate have made timely contribution to an interpretation of the expansive Creation. Each architect contributed in the particular measure decreed by fate and time. Copernicus, Halley,[3] Kepler,[4] Galileo, Huygens,[5] Newton, Herschel,

[1] Henry Wadsworth Longfellow (1807-1882) was one of the most influential American poets of the 19th century.

[2] "The Builders" was first published in his 1849 poetry collection *The Seaside and the Fireside.*

[3] Edmond Halley (1656–1742) was an English astronomer, mathematician and physicist. In September 1682, he used Newton's law of universal gravitation to compute the periodicity of Halley's Comet in his 1705 *Synopsis of the Astronomy of Comets.* It was named after him upon its predicted return in 1758, which he did not live to see.

[4] Johannes Kepler (1571–1630) was a German astronomer, mathematician, astrologer, natural philosopher, and a key figure in the 17th-century Scientific Revolution, best known for his laws of planetary motion.

[5] Christiaan Huygens (1629–1695) was a Dutch mathematician, physicist, engineer, astronomer, and inventor who is regarded as a key figure in the Scientific Revolution.

Laplace,[6] and others in the lengthy roster of time's workers assisted in the perfecting of a conceptional mechanism which explained the conditions and events, *seeming or factual*, projected on life's screen by surpassing creative function.

Yet despite the best application of time's workers, the reality remained obscure, and the most precise mathematical systems failed to embrace sublime cosmic reality. It is true that their artistry developed a materialistic system, which provided plausible and acceptable explanation of the appearance of celestial things and conditions. But the mysteries of the Cosmos remained as mysterious as ever.

Through the forceful dictates of fate and time, the systems evolved accomplished no greater knowledge of the Creation's values. They only extended the spacious lawns and gardens of assumption to dignify man's prison of terrestrial isolation. The terrestrial remained a prison in spite of the architectural enterprise.

The monumental man-made mechanistic Universe has throughout the years been embellished by all manner of astronomical "findings." And, though the things and conditions comprising such "findings" were of the illusory, popular concept has attributed to them the value of creative reality.

Theory's lawns and gardens have been so enlarged during the past four hundred years that casual observers have lost sight of the fact that they obscure a terrestrial prison. The progress of the centuries has been that of enlarging and beautifying a heathen god-image which might be expected to develop godly attributes in the process.

Such being the case, the centuries of magnified glamour for the decorative mathematical formulas may have led one to believe in the reality of the mechanistic systems which disintegrate the Cosmos and isolate the Earth. The fables of

[6] Pierre-Simon, Marquis de Laplace (1749–1827) was a French polymath whose work has been instrumental in the fields of physics, astronomy, mathematics, engineering, statistics, and philosophy.

that decorative scheme have become so firmly established that they are considered to represent factual elements of the creative pattern.

Hence there may again be expressed the thoughtlessness of a certain charming but misguided lady of other years who attended the author's lecture account of celestial reality. At the lecture's close, she artlessly exclaimed, "Oh, I do not like you! You take away my stars." How could the "stars" of that dear lady, and of all the dear and charming ladies of this Universe, be taken away, except by divine decree of the sublime Creative Force which originally ordained their resplendent but beguiling placement? Such meaningless plant is akin to the unexpected utterance of one who had long prayed to be a mother and who, in observing delivery of the infant for which she had prayed, might cry out to the obstetrician, "Oh, I do not like you! You have taken away my stork. You have destroyed the value of my childhood dolls." Would one expect that mother to renounce and condemn the medium whereby the reality she prayed for was brought to light? Could she be expected to decry the living image holding reality for all the illusions that could be crammed into human consciousness?

The tangible and the real is sought from earliest childhood. Every activity is directed toward the acquisition of knowledge which discloses new facts of the immediate world in which we dwell. And who would have it otherwise? Has the beneficent light and warmth of the Sun been depreciated through acquisition of knowledge as to the manner in which that light and warmth is generated and dispensed? Has the golden sunshine diffused from our immediate sky, wherever one might dwell become less golden because recent stratosphere observations disclose that the Sun is red, rather than of golden luminosity, when observed against stratosphere darkness? Are dreams to be considered less than dream through knowledge of the causes and the possible portent

ch. ends p. 226

of dreams? Would thought be detracted from if we were to become cognizant of the precise order and movement of a single thought vibration within the human brain? Could it be possible to consider blood less than blood if and when we acquire precise knowledge of its composition, and are thereby enabled to reproduce it in laboratory endeavor?

No, dear lady, nothing has been taken away. Your "stars" will continue to shine in the six magnitudes of their original classification, according to brightness, by the ancient gentleman named Hipparchus.[7] And they will continue to be observed unto the twenty-first magnitude by the modern gentlemen with lenses who are known as astronomers. The only thing to undergo change will be adult understanding of "star" value; and the only thing to be taken away will be the purposeless illusion of yesteryear. And though your interpretation of all such points of celestial sky-light becomes more articulate, you will never be denied pleasure of the continuing illusory appearance of your little "stars" that *seem* to "wink and blink" at you, and hold stealthy rendezvous in the stillness of the night.

The so-called "stars above" will remain to all observation. But their true character will be known. And their previous "star" value will exist in a way comparable to the manner in which animals and objects existed without body proportion for the undeveloped child mind. The minds of children not old enough to have acquired a third-dimension concept of mass or body property cannot perceive the fullness of animals and objects. Hence the animal or object must be drawn without body fullness. And all efforts to reproduce the animal or object of three dimensions, length, width, and thickness, permit of nothing more than the lines showing

[7] Hipparchus (c. 190–120 BC) was a Greek geographer and mathematician regarded as the founder of trigonometry, but is most famous for his incidental discovery of the precession of the equinoxes, and is considered the greatest ancient astronomical observer and, by some, the greatest overall astronomer of antiquity.

the animal or object on a two-dimensional plane. Without concept of the body thickness of animal and objects, the child cannot express what concept does not hold. As the child grows older, it develops three-dimensional plane. Without concept of the body thickness of animals and objects, the child cannot express what concept does not hold. As the child grows older, it develops three-dimensional concept of things. It realizes that the animals and objects have body for, or fullness. Then it is able to reproduce the animal or object *as it is* rather than as it at first *seemed to be* to the undeveloped child mind.

Strange as it may seem to members of our enlightened modern society, there are entire tribes in remote and uncultivated areas of the Earth whose members are incapable of depicting objects and animals of three dimensions. They, to, are obliged to draw the animal or object without body fullness.

Thus, would one consider that the child had lost or gained through that measure of mental growth enabling it to perceive the reality of things and conditions as they exist in a world of three dimensions? Could the devoted parent or the conscientious teacher be expected to decry the child's mental development? Would the particular animal or object become less real to the advancing child intelligence? The answers are most obvious. Nothing was subtracted from the child's mind and the measure of amusement derived from drawing the animals and objects. Nor was anything taken away from the animal or object, and the drawings thereof. On the contrary, there was considerable of lasting value added for the child, for the animals and objects, and for the drawings.

Therefore, the child mind acquired the realistic value of things. In like manner will there develop general advancement through discernment of the factual value of celestial lights. In the deeper astro-mathematical endeavor, there will continue to be telescopically observed the so-called "stars"

ch. ends p. 226

of brilliancy to the twenty-first magnitude. And "star"-light intensity will continue to be observed as varying from time to time and from place to place. That will apply to the terrestrial as well as the celestial.

Such conditions will endure for the lenses. And the numerous other deceptions, for which the lenses are responsible, will not be ended as far as observation is concerned. But the brain will know the reality behind the deceptions. Celestial observation and study will be advanced through observation of terrestrial sky-light from newly acquired celestial land points of observation. But the study will continue to hold the *apparent* features of present astronomical study of the celestial. And the apparent conditions must endure despite the fact that rocket-camera photographs have proved such features to be just as apparent in terrestrial sky-light areas.

In no way will the presently *observable* celestial pattern be changed. But its multiple manifestations will be understood for what they are, rather than what they seem to be. And the mental portrait acquired of Universe reality will transcend the mechanistic vista evolved from deceptive appearances which previously obscured reality.

The intriguing cosmic arrangement will, *to observation,* continue to contain the "giants" and the "dwarfs" of astronomy's elaborate "star" cataloging. The numerous "galaxies" will persist in the telescopically observable pattern of the cosmic whole, whether observations be from the terrestrial of the celestial. But their meaning will be known. And the meaning will express something in a realm of cosmic reality where all of yesteryear's illusions-accepted-as-fact will be known as illusion. Then will better-equipped "architects of fate" accurately read the *sky-light* prints of the Master Builder's Universe construction.

The present so-called "Heavens above" will continue to hold all the current guidance expressed by astrology, for knowledge of the movement of celestial sky-light will not change the movement. And the uplifting influences will remain for men and women who believe in the value of the "positions" of their celestial light guides. The spiritual uplift and moral guidance will be the same even through the presently assumed "ascendancy" of a particular luminous celestial area is conclusively established as nothing more than the undulating motion of luminous sky gas over an unobservable celestial land mass. It is the measure of belief and the depth of faith in a condition or thing, rather than the property of the condition or thing, which develop the inspiration and the roseate[8] outlook we all require in the journey through this "vale of tears." Hence in the ultimate it makes little or no difference how the uplift and guidance is acquired.

The art of astrology will retain its "star" symbols. Their movements, real or fancied, need not be discarded. And whatever the extent of human enlightenment may be, knowledge will not detract from the favorable influences accredited to, and forthcoming from, individual actions at the times considered to be most opportune.

In another realm of terrestrial human relations, the concept of theological Heaven can endure for the religious multitude. The most skeptical cannot successfully challenge the theological premise that the unknowable infinity contains a departed spirit abode. And, in being such, it can be expected to defeat any application of abstract mathematics seeking to determine or to negate Heaven's existence. When it is fully realized that the vast astronomical resources, with unlimited scope of operation for probing the universe about us, fail in detecting and establishing realistic values of the Universe, it will become manifest that fathoming of a more elusive spirit domain is beyond the ability of astronomy. And

[8] In this context: optimistic or idealistic.

ch. ends p. 226

it would make no difference if the spirit domain were within or beyond the physical Universe.

Moreover, were such a Utopian haven to exist within the realistic Universe, and were it to be nightly viewed and measured by all of astronomy's mighty instruments, how could its identity be established? Would the spirits tell the astronomers, or would God tell them? Could the flaunted astronomical mechanics, which are proved impotent to detect celestial land mass or to differentiate between seeming and factual sky-gas motions, be expected to penetrate into and determine an eternal celestial homestead for human spirits departed? And how could it be known as such even though it might, in some inconceivable magic manner, be embraced by mortal man's instruments of detection?

Further, which of man's great instruments could be expected to determine that the spirits detected in an obscure spirit domain were in fact heavenly spirits? What could be the precise astro-mathematical formula providing the standard of measurement for spirits heavenly and spirits unheavenly? Heaven, theological Heaven, which is not the so-called "Heavens above," could be anywhere within the constructed physical Universe, so far as any abstract science is concerned.

What abstract science, or what positive science, is capable of contradicting the conjecture that on some land-mass area of the Universe whole, and an area that is not embraced by dogmatic Heaven, there now dwell human beings possessed of wings? When we consider astronomy's absurd assumptions which obscure and deny the reality and life of the Universe, what strangeness could possibly attach to the assumption that living men and women of other Universe areas are endowed with wings? There is nothing strange about it, when we consider that any number of inferior animals of so-called prehistoric times are portrayed with wings, even though they were never seen by men. Who is to determine that the age-old desire of terrestrial man to fly stemmed in its entirety from the ever present example and influence of birds in flight? Could there not have been retained within man the instinctive Knowledge of having flown at an earlier period of his development?

Further, could not the presently developed terrestrial man, prior to his terrestrial residence, have had wings suitable for a former residence somewhere on the celestial? Surely it is just as easy to ordain men with wings as to conjecture them with tails, even though tails might be considered more appropriate for some.

Further, what mortal eloquence of reasoning can convincingly deny the existence of a celestial area inhabited by, and restricted to, formless spirits that cannot be seen? As such spirit cannot be seen, human mind could not discern their presence even though terrestrial men were to trespass on such celestial area of spirit domain and move among the formless spirit residents.

Can we, of physical substance and form, *see* the radio image of substance during the period when it is transformed into energy in motion? Can we detect it before it is received and reproduced as substance image by the receiving apparatus we have constructed especially for the energy's reception and its transformation into an image of the original substance?

And, though our receiving and transforming equipment be most magnificent, can we detect, receive, and transform the energy unless there is proper reception, or *attunement*? Can we decipher the telephonic vibrations in transit and before they reach the receiver adjusted for their reception? Can we intercept the brain's functional magnetic vibrations before they are registered as waves on the recording chart of our own making? And even after their recording, can we decipher their vibrative messages in physical terms?

These forces at work are within the unquestioned realistic

ch. ends p. 226

realm of human physical expression. They represent elements of and for man, and of which man has daily experience. Yet man, as creating power behind such forces at work (with the possible exception of the brain's function), lacks complete mastery of those forces directly at hand and under man's constant supervision. Therefore, what is the possibility of scientific determination of spirit vibrations which are without conformance to any man-made recorder? And the possibility becomes more remote if we grant the astronomical distances involved to be real.

This treatment of spirit may seem to conflict with previous mention of a living person's observation of a moving luminescent spirit proceeding in the darkness away from a human body where all vital functions had just ceased. However, there can be no conflict. The spirit seen as an individual spirit must lose its individuality as it merges with all spirits in the unknown spirit world. Then it may defeat mortal ability *to see it again as the individual spirit* as it took flight from the body it had sustained for one or one hundred years. Like the individual cell which is lost to view by the ensuing multiplication of cells constructing the human body, the individual spirit must be lost to view in its mergence with the countless spirits making the eternal spirit world. After all, it was the unseen spirit which actuated the original cell to build the body. Without it, there would have been no body. And the spirit, which actuated the original cell to build the body, *remained the actuating force of that particular body until the spirit was ready to depart.*

Such condition is life. It should be manifest to all even if there were not a single religious utterance attempting description of man's eternal spirit.

However, in spite of the individual spirit's mergence with other spirits after it has performed its task in the individual body, it may at times reassert individuality and take flight from the domain of collective spirits departed. That is a very pleasing conjecture, and there is no authority to deny the possibility. In such case, the individual spirit may again be seen by selected human beings to whom the spirit manifests its presence. The following simple example may more adequately describe. As living individuals, with body and spirit, we are permitted to *see* neuron activity of the body's nervous system; it is seen through the experienced twitching of a single nerve. But we are denied *seeing the accumulation* of body neurons which comprise the body's nervous system. Hence the departing single, or individual, body spirit at the time of departure from the body may here be considered *analogous to the single nerve's observable twitching.* That individual spirit's completion of flight from the body, making for its mergence with all the spirit world, would afford it corresponding status in the unseen accumulation of neurons in the living body's nervous system. It would thereby become invulnerable to the sight of any living person.

However, even though it were obliged to remain merged with other spirits of the spirit world, it could express *unseen spirit individuality* by manifesting its spirit presence to the spirit of a particular living person. Thus would spirit manifestations, unseen, develop for the person's subconscious, which would in turn *alert consciousness to that spirits presence.* And the spirit presence, though unseen, would be most real. The living person's entire nervous system would feel it. And the effect of the living person's spirit attunement to the departed spirit's presence would penetrate to the outer layer of the person's skin.

There are many who have known such spirit attunement, and have experienced its reaction on the flesh and the skin.

Hence it should not be too difficult to discern that the greatest possible physical advance into land areas of the so-called "Heavens above" can never involve trespass on the territory of Heaven, wherever it may be. Though the

ch. ends p. 226

so-called "Heavens above" are everywhere. Heaven must always be a restricted domain where living beings are denied entrance. Were it otherwise, Heaven would cease to be Heaven.

And it is no doubt the only area where there is no necessity for the luminous sky-light to express "stars shining above." The splendor of Heaven would have to be too magnificent for detection by lenses and their lensmen, or it could not be Heaven. It would have to transcend mortal concept. And it does.

Fifty long and tumultuous years ago, in that burdenless childhood of folklore and fables holding the enchantment of "Twinkle, twinkle, little star, how I wonder what you are," a sensitive child asked his beautiful First Lady of Life, "Mother, how far is the sky?" And the beautiful First Lady, to whom this book is appropriately dedicated, responded, "Darling, the sky is millions of miles away."

Memory of her loving response provokes the question: Can anyone believe that the measure of enchantment held in childish vista of an unknown sky a million miles away can compare with the fascination held in adult knowledge of the sky's propinquity at ten miles? Can the enchantment of distance, which served childhood, compare with adult comprehension of the sky's godly ordained purpose of providing unfailing protection for all life and vegetation on land underlying that sky throughout the Universe whole? What possible loss could the child sustain through realization that the million-mile distance was untrue, and that the appearance of great distance to the sky was an illusion?

Nothing could be taken away, because nothing real had existed. And, in this particular instance, considerable is gained through understanding of the sky's propinquity and its marvelous lifesaving purpose and function.

By the same token, what loss could be sustained from understanding that the myriad celestial lights are of the same gaseous content as the terrestrial sky, and that they express the same degree of brilliancy, and that they perform the same motions as our terrestrial sky's luminous outer surface? And who could be hurt through knowledge that the light from terrestrial sky area must express to celestial observers the same "Heavens above" which celestial lights present to observers dwelling on this terrestrial area? Though every living person possessed complete understanding of celestial reality, would not such luminous celestial areas continue to transmit the present illusory "star" messages?

We must not lose sight of the fact that "up" is always relative. "Up" is everywhere. Hence present residents of the terrestrial will in future years dwell on land underlying what is now considered a "star." Then, in looking "up," or out, from the celestial land area, they will observe terrestrial sky areas as "stars," and "planets."

And would not the future residents of celestial areas speak of the collective luminous terrestrial sky areas as "the Heavens above"? The appearances, and the description of such appearances, will continue to be the same in spite of the fact that knowledge of the illusion will be positive. It will be known that every point of terrestrial sky-light is only deceptively globular, and therefore only apparently isolated. Hence the words of illusion will endure though knowledge is had that they applied only to the illusory. They will have extended life in the manner that the "Fable of the Stork" is afforded expression by adults who know that the stork's delivery of babies is pure fiction.

Does not adult intelligence enjoy the most far-fetched fiction and the most impossible, but temporarily intriguing, cinema productions, even though complete awareness is had that conditions described by books or cinemas area beyond the bounds of reality? Hence would the utmost knowledge of celestial values cause the "stars," as they are now seen, to appear less than what they now appear to be? Would they

ch. ends p. 226

not hold greater value as known "star" illusions than as the unknown illusions of the centuries?

The "Moon" would not be less "Moon" were it universally known that its area of luminosity, greater than the luminosity of other celestial areas, is but a reflection of the Sun at various angles at different periods. And it will not detract from the "Moon" and its purpose when it is known that the reflection is not cast upon an isolated "Moon" body much closer to the Earth than other celestial areas, but that the reflection is in fact cast upon an area of the luminous connected celestial sky. Would not the "Moon" continue to shine? And would it not continue to inspire all the poetic description of yore? Would not the "harvest Moon" of tomorrow as of yesteryear parade in regal splendor along its full-dress course of autumn nights? And would it not bring to pleasing fruition the bountiful crops, and other joys of "harvest Moon" and harvest nights? Would not the symbolic "crescent Moon" persist, and merit all the time-worn description of oriental intrigue? And how dismal the soul would be of one who could not be transported on the "crescent Moon" to faraway desert sands and tents where nearby harem's passions gild the oriental "crescent Moon" with tone of fiery red.

Would not all that apply, whatever the "Moon" may be in a world of reality? And, in that world of reality, the "Moon" is very definitely not an isolated body.

The author, who fifty years ago questioned his mother, recently directed the same question to a youth who was intently observing the nightly drama of celestial sky-light. He asked, "Son, how far away do you think the sky is?" And the youth responded, "The sky is gillions and gillions of miles away."

"Gillions and gillions of miles away." As there are no gillions of which the youth spoke, there exists no isolated "Moon body" of which older children speak. Nor do there exist anywhere in the created Universe whole the isolated "star" or "planet" bodies of which astronomers speak. They are no less conditions of a world of illusion than the sky's seeming distance to the undiscerning youth to whom the sky appeared beyond estimate of distance.

So again the question is presented: What loss could that youth have suffered when he subsequently learned that there are no gillions of anything and that the seemingly distant sky is only ten miles from the Earth's surface? Likewise, what loss could be known by all the Earth's children through extension of knowledge that "stars" are *deceptively appearing globular and isolated areas of a continuous and unbroken luminous outer sky surface?*

And would there not develop a measure of spiritual uplift from knowledge that such sky protectively covers every foot of the celestial land in the same manner as it protects all terrestrial land and life? And what too would be sustained by learning that the universal sky-light, of varying brilliancy, only *seems to twinkle or blink* for the substantial reasons described in previous chapters?

Despite the acquisition of such corrective knowledge, today's children grown will in tomorrow's expanding horizons continue to look out from terrestrial positions to view the resplendent so-called "Heavens above." And they, too, will mention their favorite "twinkling stars." And their view, and the description of that view, will remain though knowledge will then be had that former terrestrial residents are *living on the land mass underlying the celestial sky area* to be seen from terrestrial observation as a "twinkling star."

Therefore, the undiscerning lady lecture attendant may take comfort in the knowledge that nobody and no known force can take away her "stars." The astrologers and their followers, and all zealous "star"-gazers everywhere, may know that their "stars" will endure as long as the Universe and its life continue.

If the Creative Force arranging the universal sky-light,

ch. ends p. 226

which permits "star" patterns to be seen for the reasons they are seen, were to cause discontinuance of the sky and its light, there could then be no mortal eyes to behold that the "stars" were gone. For without protective celestial and terrestrial sky density to produce the light which provides the "star" appearance, there would then cease to be any semblance of life on Earth or on the Universe about us.

For astronomy and its elaborate mechanistic system, the North "Star," and every presently charted celestial sky-light point comprising the astronomer's "star charts," will *remain to observation*. And they will suffer no disturbance whatever other than that of having added to them, through human understanding, their natural underlying and long-denied land mass. And it will then be understood that the underlying land mass is productive of abundant vegetation, and that it sustains human and other animal life.

No, the "stars" are not to be taken away by man's immediate conquest of celestial land areas which the so-called "stars," as areas of celestial sky-light, so competently protect and hide. The religions and their devout members will continue to retain their luminous symbols as "the Star of David," or "the Star of Bethlehem." The presently observed celestial and terrestrial sky-light appearances will endure as long as the protective universal sky remains an aspect of God's great miracle, and serves as that Master Builder's Universe roof.

The past quarter of a century's naval research and exploration has proved the disclosures first made in the presence of the Boston cardinal of 1927. It confirms that the so-called "Heavens above" are to be observed from any location of the Universal whole. However, though a thousand polar expeditions penetrate a million miles and more into the interior of the "Heavens above," there will be no disruption of the presently observable celestial pattern. The observations will forever remain as they are.

But journeys into the universe about us will provide

belated knowledge of cosmic reality. And that knowledge will inspire a greater faith in the Master Builder responsible for the Universe structure. Then will it be known that the unique Master Builder always deals in realistic force and substance which permit no place for the cosmic phantoms of astro-mathematical deduction.

The kingdom of the "Heavens above," though not of Heaven, is at hand, where it has always been. We just didn't know it. And the now clearly defined and most convenient land courses into the realistic celestial lands extend *straight ahead* from either supposed end of the known Earth. They are the *land highways* discovered beyond the South Pole point of theory on the memorable date of December 12, 1928, and beyond the North Pole point of theory in February 1947.

During the period of this book's compilation, Rear Admiral Richard Evelyn Byrd publicly announced his intention to return for exploration of the millions of square miles of land embraced by the 1928 estimate of a five-thousand-mile land extent beyond the South Pole point. Since that announcement a U.S. Naval air unit penetrated *2,200 miles* of the land extent estimated. Yet only a brief mention was made of that surpassing accomplishment of January 13, 1956.[9]

As previously explained, it should be realized that the 1928 estimate of land extent constitutes only an elementary evaluation. The five thousand miles is the greatest possible length estimate until a new estimating point is established at the five-thousand-mile location. Then another five-thousand-mile estimate of land length will be made. And that process of estimating and penetrating to the estimated length will continue for any number of years depending upon the speed of penetration into worlds beyond the Poles.

But by the time naval polar expedition of the United States and other nations reach the end of that five-thousand-mile

[9] Widely reported on January 9, 1956 (see following page for examples).

ch. ends p. 226

estimated extent, there will be found the race of men who are presently unknown to this Earth. They also have lacked knowledge of their land's extension into the terrestrial area, and they have made no attempt to penetrate the forbidding ice and storm barrier of the terrestrial's southern polar area.

Their relation to terrestrial inhabitants corresponds to our pioneering European ancestors' relation to the American Indian. The American Indian of the fifteenth century was also without knowledge that the water of the Atlantic and Pacific oceans was the course to another world. The American Indian was as ignorant of the existing "Old World" as our European ancestors were of the Indian's "New World." Moreover, the seeming meeting of the sky with the water was as real for the "New World" Indian as it was for the fifteenth-century European. Hence the Indian could not have been expected to attempt penetration into a land which was beyond his concept. And he, too, was afraid of "falling over the edge" of the Earth and being lost in space.

The international polar expeditions of 1957–58 may have penetrated to the estimated five-thousand-mile extent beyond the South Pole. As progress is continued beyond that point there will be found the numerous racial groups characteristic of this terrestrial area's population. White men will dwell in one area; black men will live in another area. Yellow men will greet explorers in a land area farther beyond. Brown and copper-colored men will be found to inhabit other areas.

All the known changes in climatic conditions common to terrestrial areas will be found to prevail throughout the land areas containing the various racial groups of worlds beyond the Poles.

And every area of the land beyond is a spacious highway of the so-called "Heavens above." For, as the illustrative fly-wheel Universe conveyed, the lowest angle in progress beyond either terrestrial Pole point bears the relation of

Admiral Byrd Joins In Plane Flight
Over South Pole In Antarctic Expedition

By SAUL PETT

MC MURDO SOUND IN THE ANTARCTICA, Jan. 9.— (AP) — Rear Admiral Richard E. Byrd has made his first flight of the current Antarctic expedition over the South Pole and the unexplored heartland area of the Antarctic Circle.

It was the veteran explorer's third flight over the geographic pole. But previous approaches have been from coasts of Antarctica closest to it, rather than over the less accessible approximate center of the continent.

The flight yesterday capped a busy six days for the air arm of Operation Deep Freeze. In all, its Navy fliers have flown over the geographic pole twice and once over the magnetic pole.

In the Antarctic quadrant facing Australia they have seen about a million square miles—some 750,000 of them never before seen by man, according to an estimate by Commander Gordon Ebbe, air operations director. They have discovered two mountain ranges and plateaus as high as 13,000 feet.

They twice have flown over areas earmarked for the Russians under the International Geophysical Year program but have not seen the Soviet party. Moscow radio said Saturday a Russian team had landed on the Antarctic ice cap and had sent out exploratory ski groups.

While all this is in the name of science, observers point out the flights can also be of great significance when and if the United States lays claim to Antarctic territory. Thus far it has made no claims and recognizes none.

"Never has so much ground been covered in so short a time," said Byrd, the director of U. S. Antarctic programs. "If the expedition does nothing else, it already has proven noteworthy and all credit is due to Commander Ebbe and his men."

Yesterday's flight took 11 hours, the round trip covering about 2,200 miles. The 67-year-old admiral said the trip was a "little tiring" but that he felt fine.

Also aboard the Navy version of a DC4 aircraft was Dr. Paul Siple, the expedition's scientific director who will head the polar base next winter. The plane was flown by Lt. Commander Henry P. Jorda of San Francisco and Lt. Commander John Donovan of Buffalo with a six-man crew.

When the plane took off it was not scheduled to fly to the geographic pole, but rather to a distant point in the interior at latitude 82 degrees south, longitude 40 degrees 50 minutes. But a heavy overcast, icing and threatening weather stopped the plane 360 miles short of its original destination and the fliers turned south and headed for the geographic pole.

At the point of turn their position was 85 degrees south 90 degrees east, which the admiral described as the "area of the pole of inaccessibility."

The plane overshot the geographic pole the first time by 40 miles. The party returned, made sure they were over the exact point and circled three times—each circle taking them around the world in two minutes.

Adm. Byrd Flies Over South Pole

McMurdo Sound, Antarctica (P) Rear Adm. Richard E. Byrd has made his first flight of the current antarctic expedition over the South Pole and the unexplored heartland area of the Antarctic Circle.

It was the veteran explorer's third flight over the geographic pole. But previous approaches have been from coasts of Antarctica closest to it, rather than over the less accessible approximate center of the continent.

The flight Sunday capped a busy six days for the air arm of Operation Deepfreeze.

Over Twice

In all, its navy fliers have flown over the geographic poles twice and once over the magnetic pole.

In the antarctic quadrant facing Australia they have seen about a million square miles—some 750,000 of them never before seen by man, according to an estimate by Cmdr. Gordon Ebbe, air operations director. They have discovered two mountain ranges and plateaus as high as 13,000 feet.

The flight took 11 hours, the round trip covering about 2,200 miles. The 67-year-old admiral said the trip was a "little tiring" but that he felt fine.

Byrd Makes His Third Flight Over South Pole

McMurdo Sound in the Antarctica—AP—Rear Adm. Richard E. Byrd has made his first flight of the current Antarctic expedition over the South Pole and the unexplored heartland area of the Antarctic Circle.

It was the veteran explorer's third flight over the geographic pole. But previous approaches have been from coasts of Antarctica closest to it, rather than over the less-accessible approximate center of the continent.

The flight Sunday capped a busy six days for the air arm of Operation Deepfreeze. In all, its Navy fliers have flown over the geographic pole twice and once over the magnetic pole.

In the Antarctic quadrant facing Australia they have seen about a million square miles—750,000 of them never before been seen by man, according to an estimate by Comdr. Gordon Ebbe, air operations director. They have discovered two mountain ranges and plateaus as high as 13,000 feet.

They twice have flown over areas earmarked for the Russians under the International Geophysical Year program but have not seen the Soviet party. Moscow radio said Saturday a Russian team had landed on the Antarctic ice cap and had sent out exploratory ski groups.

While all this is in the name of science, observers point out the flights can also be of great significance when and if the United States lays claims to Antarctic territory. Thus far it has made no claim and recognizes none.

"Never has so much ground been covered in so short a time," said Byrd, the director of U. S. Antarctic program. "If the expedition does nothing else, it already has proven noteworthy and all credit is due to Comdr. Ebbe and his men."

Sunday's flight took 11 hours, the round trip covering about 2,200 miles. The 67-year-old admiral said the trip was a "little tiring" but that he felt fine.

being "up" from terrestrial level. Study of that **Figure 1** will show that any area of the flywheel beyond the designated terrestrial Pole points must, from observation anywhere between the two Poles, appear to be "up" from the area embraced by the Poles.

Hence the discovered lands beyond the North Pole and the South Pole are not merely highways into the celestial, they are positive land areas of the celestial which makes the Universe about us. And they represent connecting land courses to the particular land areas of the "Heavens above" to be observed on the perpendicular, or directly overhead, from any land area of the terrestrial. The celestial areas having placement in the Universe whole at an angle of only 5 degrees beyond terrestrial level are as much a part of the "Heavens above" as the luminous celestial areas observed at an angle of 90 degrees. They are all connected areas of the continuous Universe whole.

The factual Universe contour, and the physical relation of the terrestrial to the celestial presents a truth stranger than the strangest fiction the minds of men have ever developed. But truth is supposed to be stranger than fiction.

13

Fulfillment of Prophecy's Endless Worlds and Mansions, and Tribes that Mark the Way

The value of yesteryear's prophetic announcements is known by subsequent developments which disclose the reality contained in the prophecy. Hence in concluding this exposition of Physical Continuity of the Universe and the modern features confirming its reality, there is fulfillment of yesteryear's dreams so long denied. In such manner is established the eternal worth of bygone prophets and their prophecies.

Thus, in an acknowledgment of ancient disclosures of other worlds, the events of this time show cosmic reality to be diametrically opposed to the presentations of the astronomical "star chart." And it is established for all who will see that from Pluto to Mercury, and from Cygnus to Centaurus,[1] the land mass underlying the continuous skylight of whatever magnitude of brightness is as dense as the land on which our terrestrial civilization is built. Throughout the entire celestial realm that condition applies. From

---

[1] A large constellation in the southern sky, named after the mythical creature, the centaur (a creature that is half human, half horse).

Phoenix[2] to Cepheus[3] and Lupus,[4] and from Indus[5] through the celestial areas of Delphinus[6] and Polaris,[7] there is evidenced the flashing facets of an incomparable sky-light diamond fashioned by a master hand.

The sky-light beacons, named "stars," guide the course of mariners on the swelling ocean's play. And they direct the lonely desert pilgrim who has faltered in his way.

Throughout the Creator's realistic Universe structure, the lights speed limitless messages of hope and inspiration as they dutifully weave a million luminous shrines for astrological faithfuls. What difference does it make, to one who hopes, if the sky-light areas are named "stars?" The beacons and the shrines are each and every one just patches of God's magnificent and protective sky-light which glows and fades from time to time and from place to place.

And, in spite of the illusions they present and the delusions they impose, who could conceive of greater perfection for the divine expression? Could the lights' measure of guidance be considered less through advancement of knowledge concerning their creatively realistic foundation as areas of protective sky? Could the hope and the ambitions of astrology's ardent adherents be diminished through discernment of the eternal foundation and the factual expressions of their shrines? Could it detract from the measure of spiritual uplift for the religiously devout to know that the light which shone over Bethlehem was of the nature of all celestial and terrestrial sky-light? Would not the very intensity of that

[2] A minor constellation in the southern sky.

[3] A constellation in the deep northern sky, named after Cepheus, a king of Aethiopia in Greek mythology.

[4] A constellation of the mid-southern Sky, whose name is Latin for wolf.

[5] A constellation in the southern sky. The English translation of its name is generally given as the Indian.

[6] A small constellation in the northern sky, whose name is the Latin version for dolphin in Greek.

[7] A star in the northern circumpolar constellation of Ursa Minor, commonly called the North Star or Pole Star as it lies less than 1° away from the north celestial pole, making it the current northern pole star.

light over Bethlehem was of the nature of all celestial and terrestrial sky-light? Would not the very intensity of that light over Bethlehem proclaim the superiority of the Infant whose arrival it announced? And would His magnificence be less if the light was known as sky-light or as a "star?"

Moreover, how could the light be considered more purposeful through the designation "star" when "star" has been proved to be in the category of the illusory? That truth was not known when Christ was born. "A rose by any other name would smell as sweet."[8] And the intensified brilliancy of any sky-light area would be just as bright and as purposeful by any other name than "star."

The illusion-based framework of astronomy prescribes "star chart" designations for luminous celestial sky areas as "stars" of varying brightness. And the measure of brilliancy extends from that of the first magnitude to the light-diminishing point of the twenty-first magnitude, and fainter. But that which is prescribed by astronomy represents in a Universe of reality the varying and extremely purposeful *sky-light intensity*. The variations may be considered as follows: Is the sky gas jet turned high or low? Is there a fifty-watt bulb or a five-hundred-watt bulb burning at the celestial point of our immediate observation?

Astronomical "planets," "star clusters," "double stars," "galaxies," "nebulae," or "the Milky Way" are additional aspects of the infinite celestial sky-light which extends over celestial land and water areas. And the sky and its light exist even though the vagueness of light over some celestial land and water areas defies telescopic detection. The identical variation of celestial sky-light brilliancy, now proved to apply to our terrestrial sky, would impel celestial astronomers to provide the same identifying labels of "star," "star cluster," or "Milky Way" to luminous areas of our terrestrial sky. It is no longer a secret that terrestrial sky-light areas present

[8] A line from Scene II, Act II of Shakespeare's *Romeo and Juliet.*

ch. ends p. 240

to inhabitants of celestial land areas all that which celestial sky-light areas present to observation from terrestrial land locations.

And, lest it be forgotten, the celestians must look "up," or out, from their land positions to observe the "Heavens above" presented by terrestrial sky-light areas, even as terrestrial inhabitants look "up," or out, to view "the Heavens above" presented by celestial sky-light areas.

The sky-light presentation can never change while the Universe and its life endure. From the distant and unknown hour of man's terrestrial arrival, the Creation's lights have mystified. The colorful high priests of ancient pagan ritual and then the sages and prophets of expanding civilization, wondered about the luminous splendor of celestial sky-light areas comprising our so-called "Heavens above." Some were gifted with an inner sight which enabled them to envisage other worlds of godly ordination beyond this meager terrestrial area. And their attunement with the sublime Creative Element inspired eloquent utterances of other worlds. Then vague record of their extraordinary disclosures was made on stone and parchment. And then, alas, the import of their disclosures was made obscure.

Their dictums did not represent the flaunting of shallow and boisterous egotism. They reflected pure ego linked to the unfathomable Prima Causa. Their attunement with first Cause, or God, endowed them with clearest perception of the Universe structure. Know and name that attunement as one will — a spark of divinity, divine revelation, perception, intuition, inspiration, cosmic consciousness, or whatever may please the individual fancy — the incontrovertible fact is that along the line of human march there has been from time to time the humble mortal conveyors of shining fragments of truth absolute. And that truth was so articulate that average human attempts at interpretation rendered it inarticulate. It was like a blinding light which made seeing impossible.

They of such extraordinary endowment were noble but wretchedly burdened souls. For they were designed as mediums through which tiny portions of realistic creative development were to be disclosed for the uplift and growth of mankind. Alas! That arrangement by Divine Will was not to be imposed without resentment by the multitude at the time and place of disclosure. They feared the intrusion by an unknown purveyor of so unknown a product as creative truth. Hence they whose strange inner sight permitted them to perceive beyond the ability of their brethren were never welcomed for the richness of their disclosures. On the contrary, they were viewed with alarm as some strange malady come to plague mankind.

Thus did the normal but nonetheless unwholesome fear of the unknown demand that "in a community of blind men, he who has sight must be destroyed." And destroyed they were, with hemlock drink, with crucifixion, and with other more advanced forms of assassination.

Therefore, fateful, complex, and confusing have been the attempts to interpret the Universe of reality. But the attempts have persisted since that hour of divine revelation when the soul of the ancient prophet Moses attuned to the voiceless decree of other worlds ordained from the beginning. And that decree's uplifting message of promise was interpreted through the voice of Moses to the poor in spirit of his particular time and place: "There are other worlds fashioned as this earth."[9]

Yet who among the tribes of that time and place was capable of fathoming the meaning in words which were of

[9] Likely a reference to Moses 1:33 and 1:35 in the Pearl of Great Price, part of the canonical Standard Works of The Church of Jesus Christ of Latter-day Saints, wherein God states, "And worlds without number have I created..." and "For behold, there are many worlds that have passed away by the word of my power. And there are many that now stand, and innumerable are they unto man..."

ch. ends p. 240

utmost clarity to Moses? Who of that desolate era could have been expected to place credence in the profound message Moses had received? Could the unknowing multitude of that time and place tune in, as did Moses, to the tone of creative development so extravagantly rich and fine as to be lost to average attunement?

There were, however, among the multitude a few bold souls who, though failing to grasp the import of the prophet's message, fearfully repeated the message. And the repetition caused vague record of the prophet's words to be carried along the corridors of time.

But the All-Knowing could not be defeated. He disclosed to the immortal Christus the secret of His vast Universe construction. And the Christus, with magnificent parable, vainly reiterated the earlier pronouncement of other worlds like unto this Earth. "In my Father's house are many mansions.[10] He who truly seeks will find."[11]

Again the inspiring and guiding pronouncement of revelation proved to be too profound for acceptance. Though it was never to be forgotten, it was never believed. And the Christly offer of "many mansions" was ridiculed by the scribes and the Pharisees who would not see. Their misinterpretations of Christly parable made "our Father's house," the Universe whole, a shambles of vague conjecture opposed to Christly dictum. And for nearly two thousand years access to any land area of the universe about us has been denied to terrestrial inhabitants.

At a later time and place in the advance of civilization, the meaning of Christly parable was rendered more obscure through professional and commercialized observation and abstract figuring of the Universe. Hence Christ's lofty parable which embraced creative reality was considered to have application only to the ideal of Nirvana, Utopia, and Paradise.

[10] John 14:2
[11] Matthew 7:8

Popular misconception, given form by dictates of abstract theory, held that the "many mansions" implied nothing more important than the conditioning of minds during this stage of human existence.

And the profound truth of Universe structure was supplanted by fiction evolved from hypotheses based on the illusory. That fiction, masquerading as fact, was capable of projecting a severely imposing Universe structure. But the projection of illusion as fact represented a foundationless "Father's house," the Universe whole, diametrically opposed to creative origin and Christly disclosure.

There is no record that Christ or Moses explained the reasons for the many worlds of their disclosure. Nor did they describe the land course into such worlds. But it is reasonable to conclude that Christ would have provided adequate explanation if He had survived the multitude's fear and hatred of unknown arbiters of land beyond the Earth.

That land beyond was unknown to the scribes and the Pharisees of Christ's time. Later the Koran described the conjectured extremities of the Earth as "lands of eternal darkness."[12] Hence they were fearful areas leading into Hell, and Christ's message of intended inspiration, for the theorists as well as the multitude, served only to accentuate their fear.

Now, 3,300 years after the disclosure by Moses and nearly 2,000 years since Christ spoke of many inhabited Universe areas like the Earth, there is blazoned a United Press dispatch under date of April 25, 1955. "Russian scientists to drive tractor over the surface of the Moon."[13] Fantastic? Such words apply only insofar as the new procedure, invention or

[12] Likely an out-of-context reference to Surah An-Nur (24:40), which states: "...the darkness in a deep sea, covered by waves upon waves, topped by dark clouds. Darkness upon darkness!"

[13] Our research revealed that the prevailing headline for that specific day referred to a "tank" instead of a "tractor." [See clipping next page: Russians Plan to Explore Moon with Robot Tank. (1955, April 25). *The Pomona Progress Bulletin* 58(21). A1.]

ch. ends p. 240

discovery, must be considered unreal because of its newness. Today's broad outlook should rob the plan of any element of fantasy which the narrow outlook of 1,900 years ago, or of only thirty years ago, might have demanded.

'LUNAR EXPEDITION'
Russians Plan To Explore
Moon With Robot Tank

LONDON (P)—Moscow Radio said today the Soviet Union is considering a plan to explore the moon with a tank remotely controled by radio. It predicted lunar trips by human beings would follow in a year or two.

A broadcast said the tank, carried to the moon by rocket ship, would be fitted with powerful television equipment to transmit pictures of the lunar landscape back to the Earth.

"Designing of the apparatus required at both ends its quite a feasible engineering job." the broadcast declared. "Moreover, the use of a tank as the first explorer of the moon will greatly simplify construction of the rocket ship. A tank can bear much greater temperature and pressure than a human being. Furthermore a space ship carrying a tank does not have to be protected against collisions with meteorites nor does a tank have to be brought back to Earth."

The broadcast said the 240,000-mile trip would be made in two stages. First, the space ship would be taken several hundred miles into the sky by rocket tugboats. Its own rocket engine would then be turned on by remote radio control to put it into an orbit around the Earth. Rockets would next be sent up to refuel the space ship for the trip the rest of the way to the moon.

"After such a flight it would be possible to consider sending a space ship with men on board." the broadcast said. The radio operated tank will choose a suitable landing ground on the moon for the passenger rocket. It is believed that no more than a year or two after the first flight of the robot ship will be required to prepare the first flight of a passenger ship."

The Soviet Union said last week it has formed a team of top scientists to devise a man made satellite which would circle the Earth.

It will be shown that the "surface of the Moon" is in fact a land area of the "many worlds fashioned like this Earth" of which Moses spoke. It will be proved that the "surface of the Moon" is a land area of the "many mansions" which Christ's parable mentioned. Technical divisions of the United States government have already publicly announced that, if occasion require, they could put a man on the "surface of the Moon."

Something has been written about the "Moon" in a previous chapter. Much more can be written. The "Moon" has always befuddled astronomers and their associated theorists. It does not fit into the man-made mechanistic pattern of the Universe. It continues to present itself as a celestial riddle because theorists mistakenly persist in considering it an isolated "body" remote from other celestial sky-light areas, whereas the "Moon" represents celestial sky area where solar reflection, at varying angles during our calendar month, accentuates the natural sky-light of celestial areas *in the reflection's course.* That course is dictated by the Sun's movement. Hence it is the reflection at different angles which produces for terrestrial inhabitants the spectacle commonly known as "phases of the Moon."

Such condition has lacked adequate explanation for many centuries. And it must forever be without explanation if we continue mistakenly to construe the Moon light as indicative of an isolated "body." The Moon of our observation is most definitely not a "body" of any nature, unless we wish to consider it a body of celestial sky-light holding the additional light of solar reflection.

In a realistic view of the Universe whole, it represents only an isolated celestial sky-light condition. And the isolated condition is produced by the only truly isolated body in the entire Creation: that is the Sun. Thus, through that Sun's reflection on the gaseous and moving celestial sky-light, there is developed *light shadings* conveniently described as "the man in the Moon." The shadings do not represent anything on the celestial land surface underlying the dual luminosity of natural celestial sky-light intensified by solar reflection. They are sole products of *light existing in celestial sky area over the celestial land.*

Experience has shown that the so-called "man in the Moon" light shadings may be considered any of numerous formations, depending upon individual fancy, when observed from different altitudes and under varying circumstances of observation. However, and regardless of any and

ch. ends p. 240

all interpretations of what the light shadings resemble, the dark patches in that luminous celestial Moon area remain *aspects of the luminosity.* They bear no relation whatever to the celestial land underlying the luminosity.

The most obvious condition of light shading is at no time afforded consideration by the astronomer. He seeks to establish it as an aspect of the land by intensive magnification of the celestial sky-light area already magnified through solar reflection. From that intensified magnification of light is developed the numerous *light pits.* They are submitted for unwary public view as the astronomer's classical "canyons on the Moon." Most astounding!

The *light distortions* resulting from magnification of sky-light over a celestial land area known as the Moon are interpreted as land-mass formations on the land surface of that particular celestial area. Such astronomical conclusion develops in spite of the fact that the celestial land area cannot be telescopically detected through the celestial sky-light density where the light-pit "canyons" are produced. Were the land under that doubly illuminated celestial sky area completely covered with realistic canyons known to exist on some terrestrial land areas, there is no lens capable of detecting them through the active luminous sky gas.

As previously related, that luminous and active sky gas covers the entire land of the universe whole. And recent U.S. naval research has established that it likewise covers every land area of the Earth. Therefore, the Russian government, in common with any other government, can during the next two years explore the land surface underlying the light of the Moon. Such memorable accomplishment will not require "shooting up," or out, from terrestrial land areas. Nor will any fantastic speed of movement be required. The airplane speeds of our time will be sufficient.

More important to our time is the celestial land exploration accomplished to date by that government which does not publicize all its findings for the benefit of Christian nations. As this chronicle of prophesy's fulfillment was being brought to timely conclusion, an International News dispatch of April 6, 1955, dealt with celestial matters much closer to terrestrial areas than the Moon. That message, despite its seeming phantasia, was attired in the raiment[14] of realism now adequately attiring (the also once-dreamed) sky-piercing rockets, guided missiles, and atom bombs. It spoke of reality equivalent to that of the familiar electric light, refrigerator, automobile, and airplane. It told in no uncertain terms of the United States government's expedition for conquest of land areas of the universe about us. And that conquest was not to be through the conjectured manner of "shooting up," or out, from the terrestrial level:

BYRD TO CONSTRUCT NAVY BASE ON SOUTH POLE EXPEDITION

The navy announcement said that five ships, fourteen planes, a mobile construction battalion with special Antarctic equipment and a total of thirteen hundred and ninety-three officers and men, will be involved in the expedition.

Specifications for the South Pole base provide:
The expedition shall procure the necessary material and construct a *satellite base at the South Pole.*

A satellite base at the South Pole! An unprecedented expedition of airplanes, ships, and manpower was to move straight ahead over land and, if feasible, on the waterways extending beyond the South Pole point. And that expedition was to penetrate into celestial land areas which appear to be "up" from the Earth.

Popular misconception, holding to the traditional "shooting up" fallacy, may question the necessity for such a lengthy

[14] Clothing.

ch. ends p. 240

journey to the South Pole to establish a base for movement into celestial areas. That question would be kindred to the 1928 conjecture by friends of Captain Sir George Hubert Wilkins. It may be recalled that their misconception caused them to believe that Wilkins would be "drawn through space" to another "planet" if he ventured beyond the South Pole. The question would be reasonable only in the orthodox and erroneous outlook that the terrestrial were in fact isolated in accordance with assumption of theory, we would have to "shoot up" to reach celestial areas. And since there will be no "shooting up," we are not isolated from the universe.

Hence, the planned course of the United States government should at long last provide convincing evidence that the Earth is not isolated in space. And that course of movement straight ahead beyond the South Pole should make it manifest that there is no other course. If the government officials responsible for that announcement had been planning a movement other than over accredited land beyond the South Pole, it would be unreasonable to establish a "satellite base" at such a remote point. The base could more conveniently be established in Maryland, or at any other more accessible point.

It was disclosed that the world's elder explorer, Rear Admiral Richard Evelyn Byrd, was to command the government's memorable expedition into that endless land beyond the South Pole. Rear Admiral Byrd was a very practical person who knew that he did not "shoot up," or out, from the North Pole point in performance of his 1947 journey over land and water extending beyond Earth's supposed northern end. He did not contemplate a flight movement contrary to that which would transport him from his Boston home to the Navy Building in Washington, D.C. He knew that he was to move straight ahead on terrestrial level from the South Pole point.

Prior to his departure from San Francisco he delivered the

momentous radio announcement, "This is the most important expedition in the history of the world." The subsequent January 13, 1956, penetration of land beyond the Pole to an extent of 2,300 miles proved that the admiral had not been exaggerating. For the United States base at that point is the most important base this nation, or any other nation, has ever held.

Hence the now proved movement straight ahead and on the same level from either Pole point will establish terrestrial man on the land of his celestial cousins. And our celestial cousins will bear all the physical characteristics of terrestrial men and women. For, strange as it may seem and difficult of comprehension as it no doubt is for the astronomers, celestial inhabitants have the same quality and quantity of oxygen as that to which we have access at terrestrial points.

The land extending beyond both terrestrial imaginary Poles is a minute area of worlds beyond the Poles. It is an area of the worlds envisioned by the prophet Moses 3,300 years ago. It is a land area room of the "many mansions" of Christ's disclosures 1,930 years ago.

Just beyond the northern and southern polar fringes of the terrestrial continue the celestial land and waters leading throughout the Universe whole. From such polar points we may at once and at will continue journey, without "shooting up," to the "valley of the Moon," and to Mars and Jupiter, and to any other area of the Universe whole!

The so-called "Heavens above," to be observed at every angle out from the terrestrial, begin where the northern and southern terrestrial polar ice diminishes!

A seven-hour flight into land areas of the "Heavens above" was accomplished in the memorable Naval exploit of February 1947. That performance beyond the North Pole point of theory was so simple that adequate explanation would have rendered it most confusing. And it is evident that no one was capable of explaining. In that 1947 naval-task-force

ch. ends next p.

flight there was land, and water, and vegetation, under the airplane course as progress was made north from the North Pole point. If the naval force had possessed motive supplies enabling them to continue, and the equipment to provide essential bases along the route, they could have then penetrated into the celestial for 100,000 miles and more, instead of only 1,700 miles.

The 1956 naval penetration of land beyond the South Pole extended for 2,300 miles over land area of the so-called "Heavens above." Recent and planned international polar expeditions can extend as far into the universe about us as their resources will permit. There is no end to the extent of possible penetration.

The unlimited natural wealth of celestial areas extending from the terrestrial Pole points has already developed a spirit of bitter competition between nations. And it should stimulate all possible corporate exploitation. After centuries of empty conjecture, knowledge is at hand that land routes to the untold wealth of the deceivingly patterned Universe extend beyond the ice-locked passages of the North Pole and the South Pole. Continued penetration of such areas will develop discovery of presently unknown human life, and other animal forms.

Yesteryear's dread of the fearful unknown may be dispelled in the light of unprecedented modern research and discovery; for they confirm that there is no northern or southern end to the Earth. The terrestrial world is in fact "a world without end."

It is so, or I could not have told you.

Appendices

Can We Walk to Mars, Maybe Jupiter and the Other Planets and Stars?

'Road to Saturn'
Beyond South Pole,
Scientist Declares

Writer Says Byrd's
Discoveries Indicate
Universe Solid Link

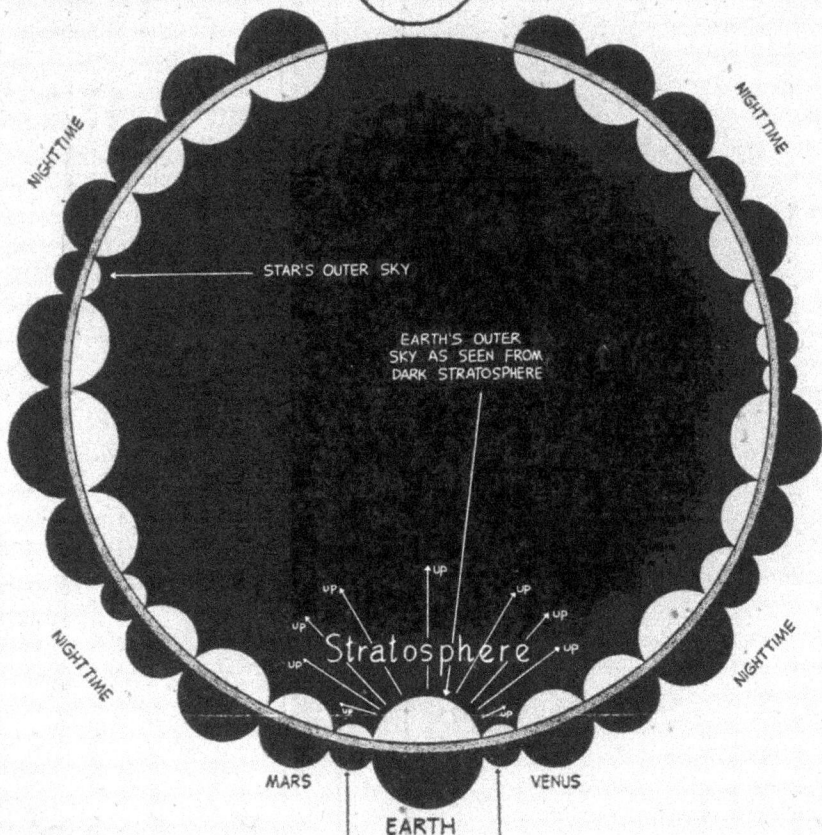

SUN

NIGHTTIME

NIGHTTIME

NIGHTTIME

NIGHTTIME

STAR'S OUTER SKY

EARTH'S OUTER SKY AS SEEN FROM DARK STRATOSPHERE

UP

Stratosphere

MARS

VENUS

EARTH

MILLIONS OF SQUARE MILES OF LAND BEING UNCOVERED BEYOND THE SOUTH POLE AND CONNECTING THE EARTH WITH THE "HEAVENS"

MILLIONS OF SQUARE MILES OF LAND TO BE UNCOVERED BEYOND THE NORTH POLE LINKING THE EARTH TO THE "HEAVENS"

In this limited drawing of the Universe the globular perspective is retained only for illustrative purposes, but there are **no isolated globes.**

All such astronomically designated parts, known as asteroids, stars and planets, regardless of all assumed magnitude or distance, are physical land areas identical with the Earth. They possess the same physical characteristics and the same physical elements.

They are all as physically connected as Boston is connected to New York, even though each observed part must deceptively appear to be isolated and, accordingly, disconnected.

Every part has oxygen to sustain life!

Every part holds some form of life!

As here illustrated, "up" appears everywhere from any position we occupy on Earth. By the same token, "up" must appear everywhere to the inhabitants of all other parts of the Universe.

Can We Walk to Mars, Maybe Jupiter and the Other Planets and Stars?

The Author: Scientist and philosopher, Francis A. Giannini has evolved a startling and amazing new theory on the structure of the universe. The first man in 500 years to challenge the accepted picture of the cosmos as outlined by Copernicus five centuries ago, Giannini claims that the recent discoveries of Adm. Byrd in the Antarctic bear out his astonishing ideas.

Born in Cambridge, Mass., in 1898, he followed a course of studies consisting of extensive research and private tutoring in philosophy, cosmology, astrophysics, and kindred subjects. Giannini's association with eminent physicists in America and Europe led to demands for his collaboration in the preparation of a complex mathematical work that was intended to go beyond the Copernican Theory. Giannini's ideas were far too unusual for his associates, however, and they continued to base their studies on the Copernican mathematical premise.

Disappointed, Giannini carried his ideas before university and scientific groups. In the light of recent discoveries of the United States Navy's Antarctic expedition, Giannini has brought his theories into the limelight once again. In survey flights over previously unknown areas, new mountain ranges, vast ocean areas and a number of lakes completely devoid of ice has been found. Even creatures reported to be 15 to 20 feet long and yellow in color were mentioned in the dispatches from Little America.

"That area beyond the Pole from Little America is the center of the unknown," Adm. Byrd has said.

As one fantastic version of the possibilities which lie beyond the poles, the Globe-Democrat presents Giannini's personally written explanation of his theory.

❧

There is a direct land path from the Antarctic region to Mars!

You can reach Venus, Saturn, or any of the other planets — provided you move in a northerly or southerly direction. But you'll never get to any of them by shooting out into space. For out there, that's all you will find — space, a vast sea of darkness, broken by cosmic-ray activity.

But you can get to any part of the universe by moving on the same level because **there are no northern or southern limitations to this Earth we dwell upon except those mathematically prescribed**. There is no end in spite of the restrictive poles that designate the ends. Poles must be remembered as no more than assumed ending places.

We cannot shoot up to Mars! And the Martians, for such entities do exist, can never hope to shoot to the Earth. No shooting up is necessary to reach any part of the universe. Our simplest and most direct route to every part of the cosmos lies straight ahead from the Arctic and the Antarctic regions. From these two points, where the 500-year-old theory of Copernicus erected the fictional North Pole and South Pole boundaries, we may journey to every point of the universe.

To reach such points, we need not resort to any means of transportation other than what is now available. We may move in exactly the same way that we now move over any portion of this Earth.

At all times during our cosmic journey we shall have

under us the same physical elements that compose this Earth. There will be land, ice, and water over the entire course from this Earth to Mars or to Jupiter or to any other part of the universe about us. Never shall we have to move through space in any way contrary to that experienced when moving by airplane from New York to San Francisco.

Not "Up" to Mars

One does not shoot up or down in moving from Boston, Massachusetts, to Hong Kong, China, and vice versa. For in spite of our mathematical concept Hong Kong is on a direct plane with Boston. Both points are on precisely the same level. And since we do not move down to China we need not move up to Mars.

The universe in its entirety, including this Earth we live on, may be conceived as an enormous doughnut or rim of a wheel as shown in the drawing on [the previous page, p. 242]. One may place the Earth at any point desired on this wheel. Then place every so-called star, planet, asteroid, and fanciful blotch of distant nebulae in their respective places in relation to the Earth. In the interest of simplicity many, many of the foregoing have been left out of the drawing. Now, consider them as appearing rounded and illuminated when observed against a dark or black background.

Thus, regardless of where the Earth is located — whether at the top of the wheel, at the bottom of the wheel, or at the center of either side — one who occupies that Earth section will see the "heavens above" in whatever direction he looks. And one who occupies any other part of the universe-wheel likewise will see the "heavens above" in whatever direction he may look. To him the Earth must appear as part of the heavens.

Therefore, whenever we move in the universe about us

ch. ends p. 252

the "heavens above" must ever appear. The "heavens above" are everywhere!

Mars, Earth on Same Plane

The system is most fascinatingly arranged! But we need never shoot up or out from this Earth to reach any of those points which deceptively appear to be "out" or "up" from the position we now occupy. As the wheel illustration shows, we shall move to all such points in the universe on the same plane and on the same level as this Earth.

The two so-called ends of this Earth, at the Arctic and the Antarctic, hold the secret of the ages. And today that secret is being unfolded by the United States Navy Antarctic Expedition. Untold millions of square miles of land wait to be discovered and explored.

In the glaring light of discoveries made by the memorable flight of Captain Sir George Hubert Wilkins, on Dec. 20, 1928, there is ample evidence that the Copernican theory — that the Earth rotates daily on its axis and that the planets revolve in orbits around the sun — has served its purpose and no longer has application to the world of things of which we are a part. Even if there were no further exploration in the Antarctic, that 1200-mile flight of Captain Wilkins changed the conception that science has held for centuries as to the geographic contour of the southern portion of this Earth.

Captain Wilkins' flight and subsequent exploration by air by Adm. Byrd unmistakably attest to the fact that there is no southern ending place for this Earth. Such flights permitted an estimate to be made of **5,000 square miles of land area which were never believed to exist**. Now, consider this: though there is a limit to the land areas we are able to estimate, today's explorers on the "under side" of the Earth are penetrating that Antarctic veil and are enabled to estimate millions of square miles of land areas which are now being uncovered.

Our Limitless Earth

And **all such land areas exist beyond the southern geographic center — beyond the imaginative South Pole**. Such land areas represent physical additions to the Earth that were always present. And that land continues not for a meager 5,000 or 10,000 miles, but its vast, unlimited extent penetrates into all of the universe about us.

As far as explorers may move north or south, from the Arctic or the Antarctic regions, they can never hope to return to our equator unless they alter their course and describe a return journey. **This Earth never has been circumnavigated north and south**. Such accomplishment is a physical impossibility for that undeniable reason in truth that unlimited land areas, extending into the universe, prohibit such enterprise.

For many years newspaper accounts have described Arctic and Antarctic explorations as journeys made "over the pole" with the result that we were led to believe that such journeys brought the explorer on the other side of the poles or, in other words, advanced him toward a return journey to the temperate zone from whence he had started. Such has never been the case. And no modern explorer would admit of such happening.

Sir Hubert Wilkins clearly described his Arctic course, **in opposition to what the press drawings conveyed**, by stating that in moving from Alaska to Spitzbergen or from Spitzbergen to Alaska he was only moving along the most northern quadrant of our supposed Earth-globe. But at no time was he moving in a course other than east to west and west to east. At no time was he moving due north, as recorded accounts would leave one to believe.

ch. ends p. 252

"Poles" Imaginary

All progress in a northerly or southerly direction, from the so-called North Pole and South Pole areas, will make for discoveries proving the Copernican Theory completely out of harmony with the factual contour and extent of this Earth and its relation with the remainder of the universe. Each mile advanced beyond the imaginative "poles" will be leading into the universe about us, which "appears up."

To achieve such noteworthy progress, however, we must completely banish from our minds the concept of the poles. They have served their purpose in their time, but today they belong in the same category as *Alice in Wonderland*[1] and the "Fable of the Stork." If we insist on clinging to the theory that the poles are the ending places of this Earth — as the boundary markings that make the Earth an isolated body — there can be no place for us to go.

Although it is true that man has peered across the far-flung horizons, he has done so only with his eyes. And from the faultiness of such optical observation there has followed the intricate and confusing maze of mathematical assumptions, resulting from our abstract calculation, which can hold no semblance of correctness. Hence, our eyes have caused us to grant existence to fantastic elements and conditions that never did exist or never could exist in the universe about us.

We have abandoned claim upon all the universe about us in the same manner as did the followers of the obsolete Ptolemaic Theory abandon claim upon half of this Earth through their denial of the physical possibility of this most factual land area we now dwell upon in this "New World." Despite the fact that figures of the Ptolemaic Theory denied the physical existence of this Western Hemisphere, it did

[1] *Alice's Adventures in Wonderland* is an 1865 English novel by English author, poet, and mathematician Lewis Carroll (1832–1898).

exist "from the beginning" in its present established relation to the "Old World" half of Earth **that we know.**

Human Eye Faulty

But it must be remembered that its existence could never have been known or proven with such a faulty instrument as the human eye. The eye definitely convinced our forefathers in the "Old World" that the sky did in fact meet the water and that such place represented the end of the physical world. It took Christopher Columbus to prove otherwise.

Therefore, is it not fitting that we take cognizance of some of the numerous pranks and deceptions of our flaunted eyesight which, though they are known to the astronomical and mathematical fraternity of our time, are accepted as absolute fact in order to hold intact the grotesque pattern of the Copernican Theory?

Do the stars twinkle? **They do not**, even though they seem to twinkle. As one moves at night over San Francisco Bay, a distance of seven miles from San Francisco to the city of Oakland, there may be observed the twinkling or shimmering movement of every electric light in Oakland. These lights appear, **but appear only**, to posses such movement because they are viewed over an area wherein the water of the bay is truly in motion. And that movement of the water between one's sensitive optic nerves and the points being observed in Oakland creates the illusion that the lights in Oakland are in motion.

Movement Illusion

Many similar cases might be cited, but this simple example proves, as does every known physical law of sight and motion, that no point of light to be observed in the universe

ch. ends p. 252

about us, regardless of its astronomical designation, **possesses any features of the movement accredited to it.**

In the 15th century Galileo observed "rounded bodies circling or ellipsing in space!" And nobody in the past 500 years has dared to express an opinion otherwise or in any way to question that magnificent fallacy which was given birth through sight's habitual pranks. **Physical necessity demands that where any point of the universe appears to be in motion it must also appear to be rounded, and therefore isolated or separated from its neighbors.**

The manner in which the human eye receives light rays is firmly established. And even elementary comprehension of a psychology of vision enables us to realize that as the human lens draws to a focal point it must cause the object under observation to take on rounded or globular proportions.

To illustrate: as one moves on a train through a tunnel that is two miles long, the entrance to the tunnel must and does describe a distinct rounded aperture when it is observed from the darkness after one has penetrated but one mile of the tunnel's length. Another illustration: as one views a long stretch of railroad tracks, there is readily observed that point where the two distinctly separate tracks must **appear to merge together.** The eye cannot have it otherwise. And it must be remembered that the eye proves nothing of the kind.

The illusions which accompany all distant observation of earthly elements by the human lens must have their counterpart in the telescopic lens in all observations of the distant heavens. **No lens, patterned as it is after the human lens, can escape the error inherent in the human lens.**

Aerial photography provides additional evidence of the optical illusions that hold such vast power to distort our understanding of the universe about us.

Air Photograph Effect

In a series of aerial photographs made by Capt. Albert W. Stevens of the Army Air Forces, one picture, taken at an altitude of seven miles, showed a 40-mile area around Dayton, Ohio.[2] In the photograph one may observe the merging or drawing together of the many Dayton streets. Were the photographs taken at even higher altitudes, with the same lens, the converging process would continue until those clearly defined streets and city blocks and buildings would represent nothing more than **a large rounded ball of pumice or granite appearing identical with the physical properties which the Moon presents.**

THE HIGHEST ALTITUDE PHOTOGRAPH EVER MADE   Official Photograph, U. S. Army Air Service

[2] Likely a reference to the aerial Stevens photograph shown above, which was taken of Dayton in 1924 from a height of 32,220 feet (6.1 miles/9.8 kilometers) and was the highest altitude a photograph had ever been taken at the time. For orientation, down is north in the photo.

ch. ends next p.

All realistic entities within that 40-mile area would be completely distorted and lost. And once that rounded deception is complete, it becomes mandatory that isolation be apparent.

The thing or area, on Earth or in the heavens about the Earth, which is drawn by the photographic or telescopic lens to a deceptive rounded body or globe must appear to be detached or isolated from all other parts in spite of the definite physical connections that exist in fact.

And as that 40-mile area would appear as a single isolated unit, so likewise would every corresponding 40-mile area of the known Earth appear as a world unto itself. The compact and physically connected Earth we dwell upon would appear to be many disconnected and isolated stars, planets, or asteroids depending upon the luminosity of the gaseous envelope surrounding each part. **A perfect reproduction of that which we now seem to see in the heavens would be shown us were we to be in the heavens and looking at the Earth**.

In other words, were we to occupy a position tonight on Mars or Jupiter or on any other so-called planet or star we would look "up" or out from that point in the same manner as we attempt to observe such point from this Earth.

The grotesque disks and globes of the Copernican Theory were never created in the universe about us except as illusion dictated. There are no isolated bodies floating around "up there" in the sublimely constructed and astoundingly bewitching pattern of the universe so artfully presented by the Master Hand.

On the Line

with Considine[1]

NEW YORK — Prof. Francis A. Giannini, philosopher and author of *Worlds Beyond the Poles*, today urged Admiral Richard E. Byrd to continue on from the South Pole to Mars. It's right next door, he says.

According to the revolutionary theories of Giannini, there is a causeway of land, ice, and water connecting the South Pole with Mars, just as he contends the North Poles connect with other "so-called heavenly bodies."

Giannini, an intense, black-eyed man from Cambridge, Massachusetts, who once served as an Antarctic chartmaker for bearded Sir Hubert Wilkins, thus returns to the public prints after an absence of 20 years. It was in 1927 that he released his theory of a connected universe, which was promptly scorned by astrologists.

"I now have absolute proof that the universe is one body," he told me today.

"I have been silent for 20 years because I lacked visual

[1] Robert "Bob" Bernard Considine (1906–1975) was an American author and insanely prolific journalist ("Considine's speed, accuracy, and concentration as a writer and his seemingly inexhaustible energy were legendary in the newspaper profession"). His commentaries were syndicated nationwide in newspapers and on radio as "On the Line with Considine." This article featuring Giannini was syndicated in early 1947.

proof of my theory. That has been provided by the camera shots taken from the tail of a German V-2 ascending over White Sands, N. M."

He produced the now familiar pictures, including the one the army says was taken at a height so great that the curvature of the earth could be seen.

"That is not a picture of the earth's curve," the professor protested. "It is a trick of the eye and the camera lens, fashioned after the eye.

"Photographs of cities taken from planes at about 30,000 feet show the tall buildings beginning to lean toward the center of the film plate. The edges appear to curl, in saucer fashion.

"Now! Let that plane soar 30 or 40 miles into the stratosphere and, if its camera could pierce the atmosphere, the area below would appear perfectly round.

"New York, Chicago, Los Angeles, and every other big city would look like separate planets, shiny and round, if a camera or human eye was lifted high enough above them.

"That most certainly wasn't the curve of the earth which the V-2's camera caught. It simply was that part of New Mexico which, from that great height, became a rounded world of its own.

"And so it is throughout our connected universe. We see these so-called planets and stars and think they are distinct bodies in space, because of the distances involved. They are part of us; we are part of them."

I was getting a bit confused by now and dragged the office-owned globe over to his chair. The sight of the globe turned the professor pale with anger.

"Take that Copernican absurdity away from me or I'll pull out my teeth!" he threatened.

"The earth, as we call it, is cylindrical, yes; but not spherical," he declared.

We decided not to scoff. Why go down in history with the intolerant boobs who laughed at Columbus' egg tricks?[2]

"The universe is shaped like a rubber tire, which revolves and exposes itself once a day to the sun. The sun, for all I know, may be pulling it through space.

"Here and there the tire, let's call it, tapers a bit and then bulges out again, perhaps like a circular link of squat sausages with heavy connecting cords. Now, think of bits of light-reflecting crystals embedded in the tire. They pick up light from the sun, and at night, when the sun isn't shining on us, we can see them.

"They appear round only because the human eye, and the telescope, are faulty windows of the brain. They are not round and not distinct. They are joined by the land, water, and ice that comprise the whole tire-shaped universe.

"It would be interesting for Admiral Byrd to go on to Mars from what he believes is the South Pole. Remember, I said 'on,' not 'up.' Talk of shooting rockets up into the air toward the moon or planets is preposterous. Would New York try to contact Chicago by rocket? Certainly not, for there is an overland route."

I mumbled something about the earth having been circumnavigated.

"Yes, from east to west," the professor agreed. "But not from north to south. Claims about the discovery of the North and South Poles are either lies or the result of childish dependence on the irresponsible instrument we call the compass.

"The earth, as we call it, simply extends on from the Poles. Byrd needn't go on as far as what we call Mars, if there are supply difficulties. We could push on 10, or 100, or 1000 miles

[2] An egg of Columbus or Columbus's egg refers to a seemingly impossible task that becomes easy once understood. The expression refers to an apocryphal story, dating from at least the 16th century, in which it is said that Christopher Columbus, having been told that finding a new trade route was impossible, challenges his critics to make an egg stand on its tip. After his challengers give up, Columbus does it himself by tapping the egg on the table to flatten its tip.

ch. ends next p.

around the universal tire, and he'll find lands, seas, and ice areas the astrologers never dreamt of."

Prof. Giannini also sees no reason why visible spots on the tire — planets, stars, satellites, we call them — are not inhabited. "The fact that they are lustrous indicates they have atmosphere enough to sustain life," he said.

Giannini doesn't like astronomers. They are too dependent on mathematics, he said, and added that mathematics can be made to prove anything. He offered to prove, on paper, that I was dead, but I declined with thanks — tomorrow being payday.

He did, however, present his theory to Harvard's eminent Dr. Harlow Shapley, leading U.S. astronomer.[3] I asked him what Shapley said.

"He said, 'Giannini, you should be in jail,'" the professor concluded, with a bitter laugh.[4]

---

[3] Harlow Shapley (1885–1972) was an American astronomer, who served as head of the Harvard College Observatory from 1921–1952, and is best known for determining the correct position of the Sun within the Milky Way Galaxy.
[4] Do you think Shapley knew?

Giannini's Theory and Byrd Findings

by Herbert E. Wilson

(Managing Editor, *The Argus*)[1]

World beyond the poles?

That's the question asked in a recent series of articles in *The Argus* by Dr. Francis A. Giannini.

He challenged the old Copernican theory of isolated planets in a disconnected universe.

He asserted the road to Mars is beyond the South Pole, and that it may be possible to walk to Mars, Venus, Jupiter, and the other planets and stars.

These articles were thought-provoking.

They appear to have stirred particularly the minds of college and high school students.

Today some of them are asking this question: "If it's possible that all the planets are linked with the Earth, why have we been taught otherwise in all our years of schooling?"

Our recommendation to these young people is: Take it easy.

Dr. Giannini's views comprise theory, not fact.

---

[1] *The Argus*, founded in Rock Island, is one of Illinois' oldest continuously published newspapers.

There are many eminent physicists who disagree with
Dr. Giannini.

But it is nevertheless true that in the light of recent discoveries by the U.S. Navy's antarctic expedition, Giannini's views
seem to have been given important and practical support.

In survey flights over previously unknown areas, new
mountain ranges, vast ocean areas, and a number of lakes
devoid of ice were found. Even creatures reported to be
large animals 15 to 20 feet long and yellow in color were
mentioned in the dispatches from Little America.

Rear Admiral Richard E. Byrd said, "That area beyond the
pole from Little America is the center of the unknown; I'd
like to see that land."

Dr. Giannini's theory is so amazing as to be almost fanciful, yet the Byrd party's findings suggests the possibility of
millions of square miles of land areas in the antarctic yet to
be discovered.

Theory 20 Years Old

Here is the background of the Giannini research:

Scientist and philosopher, Dr. Giannini is the first man in
500 years to attempt to give a greater picture of the universe.

A native of Cambridge, Massachusetts, and now a resident
of East Orange, New Jersey, he is 49 years old. His studies
have consisted of extensive research and private tutoring in
philosophy, cosmology, astrophysics, and kindred subjects.

Giannini's association with eminent physicists in America and Europe led to demands for his collaboration in the
preparation of a complex mathematical work that was
intended to go beyond the Copernican theory.

Much to Giannini's disappointment, however, his associates persisted with the Copernican mathematical premise
of isolated bodies in a disconnected universe, which was
contrary to his thinking.

Giannini conceived his intuitive portrait of the physical continuity of the universe 20 years ago. He has lectured on aspects of this theme at universities and before scientific groups.

Also, his close application to the intricate maze of mathematical calculations has enabled him to point out to leading scientists, clergyman, and engineers the fallacies of many accepted conclusions.

Modern Christopher Columbus

If Giannini's theory about the planets ever is proved correct, he will assume the stature of a modern Christopher Columbus.

Quite naturally, it appears doubtful today that proof ever will be obtained. On the other hand, who can foretell the scope of the march of science in the next five years?

In one or two quarters, *The Argus* has been criticized for publication of his series; from many other quarters has come praise.

The articles were made available to this newspaper by one of the best known and most reliable feature syndicates in America, which acquired distribution rights after careful and complete appraisal of all factors involved.

In view of the latest findings by the Byrd expedition, it seems that Dr. Giannini's thinking definitely has a place in the efforts of science to push back the horizons of man's comprehension.

Possibly his articles will stir our youth to greater interest and accomplishment in scientific fields. If so, the Giannini views — theory or fancy — will have served a constructive purpose.

FLYING SAUCERS

The Magazine of Space Conquest

WISCO

35¢

DECEMBER, 1959

**ADMIRAL BYRD'S WEIRD FLIGHT
INTO AN UNKNOWN LAND!**

**COLUMBUS WAS WRONG
THE EARTH ISN'T ROUND!**

**POLAR EXPLORATION PROVES
SAUCERS FROM THIS EARTH!**

Saucers from Earth!

Flying Saucers #13, December 1959

A Challenge to Secrecy!

Flying Saucers has amassed a large file of evidence which its editors consider unassailable, to prove that the flying saucers are native to the planet Earth; that the governments of more than one nation (if not all of them) know this to be fact; that a concerted effort is being made to learn all about them, and to explore their native land; that the facts already known are considered so important that they are the world's top secret; that the danger is so great that to offer public proof is to risk widespread panic; that public knowledge would bring public demand-for action which would topple governments both helpless and unwilling to comply; that the inherent nature of the flying saucers and their origination area is completely disruptive to political and economic status-quo.

Since the day Kenneth Arnold first brought flying saucers to wide public attention by his famous sighting,[1] one fact has been consistently brought forth by investigators — flying

---

[1] Kenneth Albert Arnold (1915–1984) was an American aviator, businessman, and politician best known for reporting what is generally considered the first widely publicized modern sighting of UFO in the United States, after claiming to have seen nine silver-colored discs flying in unison near Mount Rainier, Washington, on June 24, 1947.

saucers did not originate with that sighting, but have been with humanity for centuries if not thousands of years. Flying Saucers is the popular term for Unidentified Flying Objects (UFO), or Unidentified Aerial Phenomena (UAP). The popular "saucer" shape is only a segment of the phenomena. Properly, the important fact to consider is that intelligently controlled phenomena appear in our atmosphere. Their exact nature is a matter for conjecture, and is quite varied in configuration (to use a favorite Air Force word). It is this fact of antiquity which poses the most important single factor in analyzing the phenomena. At one stroke it eliminates contemporary earth governments as the originators of the mysterious phenomena.

Because of this antiquity, many investigators have turned from the earth to other planets, and to other solar systems. Each planet has its followers in this group of investigators, and such bodies as Venus, Mars, and Saturn are favorites with the so-called contactees. We are not at all concerned in this symposium of evidence with the contactee — a phenomenon as "unidentified" as UFO themselves. It may not even be remotely related. However, chief among the advocates of interplanetary origin is Major Donald E. Keyhoe, whose efforts are entirely directed toward collecting evidence that will serve to advance this theory.[2] The interplanetary theorist has a large following, and is perhaps the only theorist that will even be considered by scientific men such as astronomers. While it is true that there are many mysteries of an interplanetary nature, linking them with UFO demands a stretching of the evidence, and a great deal of extrapolation. It may be true that there are "configurations" on the Moon, for instance, which are used by Keyhoe to postulate "flying

[2] Donald Edward Keyhoe (1897–1988) was an American Marine Corps naval aviator, author with a focus on aviation, and tour manager of aviation pioneer Charles Lindbergh (1902–1974). In the 1950s, Keyhoe became a UFO researcher and writer, arguing that the U.S. government should conduct research into UFO matters, and should publicly release all its UFO files.

saucers" on that body. Unfortunately (for Keyhoe), this same evidence is used by contactees such as Adamski to support their contentions.[3] Actually, the Moon is remote, in reference to Unidentified Aerial Phenomena, and we must disregard it when we speak of atmospheric phenomena: events which occur within our atmosphere. Since almost all "sightings" are in-atmosphere in nature, the greatest percentage of thinking on them must be limited to the atmosphere.

Because our planet is quite well (but not completely) known, it has been easy for interplanetary theorists to prove that the strange objects are not made by any single government or group of governments on Earth. Such a vast project could not remain secret over so long a period, and also, the matter of antiquity does not allow the phenomena to be fitted into the history of existing governments.

How well-known is the Earth? Is there any area on Earth which can be regarded as a possible origin for the flying saucers? There are two, speaking in major terms, and four, speaking in more minor terminology. The two major areas, in order of importance, are Antarctica and the Arctic. The South Polar continent, and the North Polar area. We speak of the North Polar area because exploration made public to date indicates there is no land, but that it is an ocean, frozen over with ice, under which exploration by submarine is being carried on. The two minor areas are South America's Mato Grosso[4] and Asia's Tibetan Highlands.[5]

Could the flying saucers come from any of these areas? We can largely eliminate the Mato Grosso and the Tibetan

[3] George Adamski (1891–1965) was a Polish-American author who became widely known after he displayed numerous photographs in the 1940s and 50s that he said were of alien spacecraft, claimed to have met with friendly Nordic alien or "Space Brothers," and claimed to have taken flights with them to the Moon and other planets.

[4] A high plateau region of southwestern Brazil divided into two states: Mato Grosso and Mato Grosso do Sul.

[5] The Tibetan Plateau is a vast elevated plateau located at the intersection of Central, South, and East Asia.

ch. ends p. 287

Highlands; firstly because of the enormous numbers of the UFO, and secondly because these areas are not entirely unexplored, and can be flown over almost at will. Evidence is lacking in both these areas. Negative evidence, however, does exist in some measure, sufficiently to cause theorists to discard both areas, except in a minor way. At most, either or both Mato Grosso and Tibetan Highlands, can be suspected to be "bases" or something on the order of "way stations."

What about the North Pole? Explorers say it is entirely oceanic in nature, covered with ice which sometimes melts in part, and in many areas is quite thin at all times. The depth of the ocean beneath this ice varies from some 24 fathoms[6] to several miles. Flights have been made to and across the North Pole. Submarines, notably *Nautilus*[7] and *Skate*,[8] have traveled to the Pole and returned, crossing from one side to the other (Point Barrow[9] to Spitzbergen). Apparently the sort of base necessary for the UFO mystery in its entirety does not exist in the North Polar regions.

What about the South Pole? Here we have a continent quite as large as North and South America combined, insofar as land mass is concerned. At least one large area (40,000 square miles) is known to experience 100% melting during the summer, and even in winter possesses warm water lakes (from warm springs, geysers, etc). This area is under control of the Russians, who have a permanent base there. Expeditions from both Little America and from the British zone of

---

[6] A fathom is a unit of length equal to six feet (approximately 1.8 meters), so 24 fathoms would equal 144 feet or 43.8 meters.

[7] USS *Nautilus* (SSN-571) was the world's first operational nuclear-powered submarine and became the first submarine to complete a submerged transit of the North Pole on August 3, 1958.

[8] USS *Skate* (SSN-578) was the third submarine of the United States Navy named for the skate (a type of ray), the first to make a completely submerged trans-Atlantic crossing, the second to reach the North Pole, and the first to surface there.

[9] Point Barrow or Nuvuk, on the Arctic coast of Alaska, is the northernmost point of all the territory of the United States, situated 1,122 nautical miles south of the North Pole.

exploration, have reached the South Pole. Expeditions have also reached the South Magnetic Pole. This is a distinction it is necessary to stress, due to the strange fact that the South Magnetic Pole is actually 2300 miles distant from the South Geographic Pole.[10] It is a fact that a tremendous land area exists in the South Pole Continental Area which is unexplored and which constitutes a large blank on the map of the Earth.

Let us consider the North Pole first, and discover what we know about it. What are the facts about the "top" of the Earth?

First, it is surrounded on all sides by known areas of land. Siberia, Spitzbergen, Alaska, Canada, Finland, Norway, Greenland, Iceland. The northern shores of these lands border on the Arctic Ocean, in the virtual center of which both the geographic and magnetic poles exist. These two poles are separated by less than 200 miles, and one of them, the magnetic pole, is known to "wander" somewhat.[11]

The North Pole has been reached by a number of expeditions. The latest we know of are the exploits of the *Nautilus* and *Skate,* both atomic submarines which traversed the entire extent of the Arctic Ocean beneath the ice, making the Pole itself (magnetic) a stopping point. On the surface of things, it can be said that the North Polar Area is fairly well explored. In addition to our submarine explorations, the Russians have also traversed the Arctic Ocean. They have even established magnetic "bases," navigational aids which they have planted along Alaskan and Canadian shores, so that rocket-launching atomic bomb submarines can proceed

[10] The south magnetic pole is constantly shifting due to changes in Earth's magnetic field and, due to polar drift, the pole is moving northwest approximately 6 to 9 miles (10 to 15 kilometers) per year. As of 2020, its distance from the actual Geographic South Pole was approximately 1,780 miles (2,860 kilometers).

[11] In 2009, while still situated within the Canadian Arctic, the north magnetic pole was calculated as moving toward Russia at between 34 and 37 miles (55 and 60 kilometers) per year. As of 2021, the pole is projected to have moved beyond the Canadian Arctic.

ch. ends p. 287

swiftly to a prearranged launching site, and fire rockets on prearranged courses. American submarines have been busily (we hope) moving these navigational aids to new sites which throw off the prearranged calculations, thus making them worthless.

But there is an area of doubt which *Flying Saucers* intends to explore, and to present as the first of its bits of evidence which point to what may well be the best-kept secret in history. In order to do so, we must go back to 1947. In February of that year, Admiral Richard E. Byrd, the one man who has done the most to make the North Pole a known area, made the following statement: "I'd like to see that land beyond the Pole. That area beyond the Pole is the center of the great unknown."

Millions of people read his statement in their daily newspapers. And millions thrilled to the Admiral's subsequent flight to the Pole and to a point 1700 miles beyond it. Millions heard the radio broadcast description of that flight, which was also published in the newspapers. Briefly, for the benefit of our readers, we will recount that flight as it progressed. When the plane took off from its Arctic base, it proceeded straight north to the Pole. From that point, it flew on a total of 1700 miles beyond the Pole, and then retraced its course to its Arctic base. As progress was made beyond the Pole point, iceless land and lakes, mountains covered with trees, and even a monstrous animal moving through the underbrush were observed and reported via radio by the plane's occupants. For almost all of the 1700 miles the plane flew over land, mountains, trees, lakes, rivers.

What land was it? Look at your map. Calculate the distance to the Pole from all the known lands we have previously mentioned. A good portion of them are well within the 1700 mile range. But none of them are within 200 miles of the Pole, Byrd flew over no known land. He himself called it "the great unknown." And great it is, indeed! For after 1700 miles over land, he was forced by gasoline supply limit to return and he had not yet reached the end of it! He should have been well inside one of the known areas mentioned. He should have been back to "civilization." But he was not. He should have seen nothing but ice-covered ocean, or at the very most, partially open ocean. Instead he was over mountains covered with forests.

Forests!

Incredible! The northernmost limit of the timber-line is located well down into Alaska, Canada, and Siberia. North of that line no tree grows! All around the North Pole, the tree does not grow within 1700 miles of the Pole!

What have we here? We have the well-authenticated flight of Admiral Richard E. Byrd to a land beyond the Pole that he so much wanted to see, because it was the center of the unknown, the center of mystery. Apparently he had his wish gratified to the fullest, yet today, in 1959, nowhere is that mysterious land mentioned. Why? Was that 1947 flight fiction? Did all the newspapers lie? Did the radio from Byrd's plane lie?

No, Admiral Byrd did fly beyond the Pole.

Beyond?

What did the Admiral mean when he used that word? How is it possible to go "beyond" the Pole? Let us consider for a moment: Let us imagine that we are transported, by some miraculous means, to the exact point of the North Magnetic Pole. We arrive there instantaneously, not knowing from which direction we came. And all we know is that we are to proceed from the Pole to Spitzbergen. But where is Spitzbergen? Which way do we go? South, of course! But which south? All directions from the North Pole are south!

This is actually a simple navigational problem. All expeditions to the Pole, whether flown, or by submarine, or on foot, have been faced with this problem. Either they must retrace their steps, or discover which southerly direction

ch. ends p. 287

is the correct one to their destination, whatever it has been determined to be. The problem is solved by making a turn, in any direction, and proceeding approximately 20 miles. Then we stop, shoot the stars, correlate with our compass reading (which no longer points straight down, but toward the North Magnetic Pole), and plot our course on the map. Then it is a simple matter to proceed to Spitzbergen by going south.

Admiral Byrd did not follow this traditional navigational procedure: when he reached the Pole, he continued on for 1700 miles. To all intents and purposes, he continued on a northerly course, after crossing the Pole. And weirdly, it stands on the record that he succeeded, for he did see that "land beyond the Pole," which to this day, if we are to scan the records of newspapers, books, radio, television, and word of mouth, has never been revisited!

That land, on today's maps, cannot exist. But since it does, we can only conclude that today's maps are incorrect. Incomplete, and do not present a true picture of the northern hemisphere!

Having thus located a great land mass in the North, not on any map today, a land which is the center of the great unknown, which can only be construed to imply that the 1700 mile extent traversed by Byrd is only a portion of it, let us go to the South Pole and see what we can learn about it.

On April 5, 1955, the U.S. Navy announced an expedition to the South Pole. It was to be headed by Admiral Richard E. Byrd. It consisted of five ships, fourteen airplanes, special tractors, and a complement of 1393 men. The stated purpose of the expedition was as follows: "To construct a satellite base at the South Pole."

In San Francisco, on the eve of his departure, Admiral Byrd delivered a radio address in which he stated: "This is the most important expedition in the history of the world."

Let us pause a moment and pretend we are rocket men, primarily the scientist-rocket-men who are engaged in launching satellites. Our task is a troublesome one. Many failures result. Our work is a tremendously difficult task, and sometimes important rocket shoots are delayed for days by weather. Our base is a gigantic one, here at Cape Canaveral.[12] The logistics problem is enormous. The rockets themselves weigh hundreds of tons. To be asked to set up such a satellite base at the South Pole would cause us to stare in utter amazement at the official making the request. We would waste no time in informing him that he hasten immediately to his psychiatrist and retire from active service, for he has indeed "gone off his rocker." In short, a satellite base at the tip of South America in plain words, is totally ridiculous. Even a satellite tracking station at the South Pole is nothing short of idiotic. For tracking purposes, a base at the tip of South America is entirely adequate. Or on a series of ships anchored about the Antarctic Circle.

This, then, cannot be a satellite base. It must be something else. On January 13, 1956, we learn what it really is. On that date the U.S. Navy flies to a point 2300 miles beyond the South Pole. The entire distance is accomplished over land.

Once again, look at your map. Unlike the North Polar Sea, the South Polar Continent is entirely surrounded by water. And in all cases, no matter what direction you proceed from the South Pole, you pass from the continental area to a known oceanic area. You proceed hundreds of miles over water to reach a distance of 2300 miles.

Once again we have penetrated an unknown and mysterious land which does not appear on today's maps. And once again, we find no further announcement beyond the initial announcement of the achievement.

And strangest of all, we find the world's millions absorbing the announcements, and registering a complete

---

[12] A cape near the center of the state of Florida's Atlantic coast. Officially Cape Kennedy from 1963 to 1973, it is part of a region known as the Space Coast, and is the site of the Cape Canaveral Space Force Station, built in 1949.

ch. ends p. 287

blank insofar as curiosity is concerned. Nobody, hearing the announcements, or reading of them in the newspaper, bothers to get a map and check the facts! Or if they do, they only shake their heads in puzzlement, and then shrug their shoulders. If Admiral Byrd is not bothered with the apparent inconsistencies, why should they be?

Here, then, are the facts: At both poles exist unknown and vast land areas, not in the least uninhabitable, extending for distances which can only be called tremendous, because they encompass an area bigger than any known continental area! The North Polar Mystery Land seen by Byrd and his crew is at least 1700 miles across its traversed direction, and cannot be conceived to be merely a narrow strip, as the factor of coincidence in flying precisely along its longest extent is improbable. It is a land area perhaps as large as the entire United States! The land area at the South Pole, considering that the flight began 400 miles west of the Pole, and thus covers a continuous land area of 2700 miles in one direction, means a land area possibly as big as North America in addition to the known extent of the South Polar Continent, which is located north of the Pole whereas the 2300-mile land traversed by the Navy plane is "beyond" the Pole. Once more the same condition of navigation exists: progress was made to the Pole and then straight on beyond it, with the one difference that the South Geographic Pole is located 2300 miles away from the South Magnetic Pole, and it is not necessary to perform the navigational maneuver described previously. If navigating from the South Magnetic Pole, the procedure is again necessary, with differences due to the greater angle of inclination to the stars, and the possibility of navigation entirely by the stars rather than with the aid of a compass.

Let's stop here and make a statement that logically follows: the flying saucers could come from these two unknown lands "beyond the Poles." It is the opinion of the editors of

Flying Saucers that the existence of these lands cannot be disproved by anyone, considering the facts of the two expeditions which we have outlined. These facts can be checked by anyone. You have merely to read the newspapers of the day. If there is anyone who can satisfactorily explain away these two expeditions, and the statements of Admiral Richard E. Byrd concerning them, *Flying Saucers* will give him every inch of space necessary to complete his explanation.

Just for the record, let's present the actual, announcement carried by press and radio on February 5, 1956: "On January 13, members of the United States expedition accomplished a flight of 2,700 miles from the base at McMurdo Sound,[13] which is 400 miles west of the South Pole, and penetrated a land extent of 2300 miles beyond the Pole."

And on March 13, Admiral Byrd reported, upon his return from the South Pole: "The present expedition has opened up a vast new land."

Finally, in 1957, before his death, he reported it as: "That enchanted continent in the sky, land of everlasting mystery!" Which statement remains to your editors as the most mysterious of all, and almost inexplicable. "Enchanted continent in the sky . . . " Everlasting mystery, indeed!

Considering all this, is there any wonder that all the nations of the world have suddenly found the South Polar region (particularly), because of its known land area and the North Pole region so intensely interesting and important, and have launched explorations on a scale actually tremendous in scope?

And was it because of Admiral Byrd's weird flight into an unknown Polar land in 1947 that the International Geophysical Year was conceived in that year, and finally brought to

[13] A sound in Antarctica, known as the southernmost passable body of water in the world, located approximately 810 miles (1,300 kilometers) from the geographic South Pole.

ch. ends p. 287

fruition ten years later, and is actually still going on?[14] Did his flight make it suddenly imperative to discover the real nature of this planet we live on, and solve the tremendous mysteries that unexpectedly confronted us?

If you have followed us thus far, it may be that you have gone to your map or your globe and have tried to fit these mysterious lands onto the planet, and have come up with a snort and said: "These bits of evidence are all very well, but the fact remains there is nowhere physically to place these land masses. Since the space to do so is lacking, there exists a fundamental impossibility which cannot be overcome." Good boy! Don't give up your guns. Insist that we overcome this fundamental impossibility, and support our original evidence in not a few ways, but in hundreds.

Although it is obvious to your editor that the complete evidence cannot be presented in one single issue of *Flying Saucers*, we intend to continue to present it in each issue until we have covered the entire range of evidence. There is evidence, and it is totally factual. It covers the fields of astronomy, physics, chemistry, geology, anthropology, and exploration. We have spent thirteen years in amassing this wealth of facts, and we shall present them all before we have finished. The reader is therefore asked to remain with us until we have completed our presentation.

The question that most logically follows the two instances of exploration which we have outlined is whether or not other Polar Expeditions have encountered similar and confirming conditions. In order to answer this question, it will be necessary to examine the records of all North and South Pole explorations from the very first of which modern man has any knowledge. As a sub-subject, it might be interesting later on to go into legend and mythology for still further bits of confirmation, but we are concerned now only with

[14] The International Geophysical Year (IGY) was an international scientific project that lasted from July 1, 1957, to December 31, 1958.

presenting provable facts. In the presentation of these facts, we intend to draw no conclusion. They should become obvious to the reader without prompting.

To those of our readers so inclined, there must be a great deal of interest on their part in the historically famous debate on which, or both, or neither, Cook or Perry actually reached the North Pole. In the years following these expeditions, much debate went on, and even today arguments rage. Briefly, let's outline the claims of both men.

Dr. Frederick A. Cook said he reached the Pole on April 21, 1908.[15] His announcement was followed by a few days by one from Rear Admiral Robert E. Peary that he had reached the Pole on April 6, 1909.[16] Both men hurled accusations against the other, Cook even saying that Peary had appropriated some of his stores cached against his return from the Pole. Cook, in his turn failed to supply notes he said he had kept of his trip, and thereby cast doubt on his own story. The reader who is interested in the whole story should visit his library and read up on the controversy.[17]

Although Cook claims to have been the first to reach the Pole we will take Peary's claim, which has been universally recognized, and examine it. Cook's claim was discredited on one basis because the sun altitude was so low that observations of it as proof of position were worthless. It should be noted that Peary also reached the Pole in April, 15 days earlier in the season, and therefore under even more adverse solar observation conditions. His calculations therefore are

[15] Frederick Albert Cook (1865–1940) was an American explorer, physician, and ethnographer, who is most known for allegedly being the first to reach the North Pole.
[16] See footnote #4, p. 50.
[17] In December 1909, after reviewing Cook's limited records, a commission of the University of Copenhagen ruled his claim unproven. Peary's claim to have reached the North Pole has long been subject to doubt, with some polar historians believing that Peary honestly thought he had reached the pole, while others have suggested that he was guilty of deliberately exaggerating his accomplishments.

ch. ends p. 287

more suspect than Cook's. Cook, it was said, had no witnesses other than Eskimos; the same is true of Peary. Peary, however, lacked witnesses through choice, having ordered his white companions to remain behind, while he went on alone with one Eskimo companion to the Pole. Cook was doubted in his claim that he averaged 15 miles a day. Peary claimed to have made over 20. Undoubtedly the argument will never be settled. However, there is the factor regarding Peary's dash to the Pole, which, in our opinion, is quite remarkable. This factor lies in the fantastic speed with which he made his trip.

When Peary neared the 88th parallel,[18] he decided to attempt the final dash to the Pole in five days. He made 25 miles the first day; 20 on the second; 20 on the third; 25 on the fourth; 40 on the fifth. His five-day average was 26 miles. On the return trip he traveled a total of 153 miles in two days, including a halt 5 miles from the Pole to take a sounding of the ocean depth. This is an average of 76½ miles per day. His actual traveling time was approximately 19 hours per day. This is a walking speed of 4 miles per hour. Can a man walk that fast under the incredible conditions of the North Pole area, an ice-terrain described by the men of the atomic submarine *Skate* as fantastically jumbled and jagged? And yet, further south, with presumably better going, he was able to average only 20 miles per day.

We stress the distances only because the ones nearest the Pole are weirdly impossible. Only if Peary was reporting honestly would we have included such contradictory calculations which he must have known would discredit his story. Therefore we can assume that he did report honestly, and that we have a speed of travel which projects into the same mysterious area in the same "unfittable" manner as a whole vast continent fits into a space that is totally lacking. When

[18] The 88th parallel north is a circle of latitude located in the Northern Hemisphere, in the Arctic Ocean, 2 degrees south of the North Pole.

traveling over a land whose dimensions are fantastically "expanded," will we not also travel at an equally fantastically "expanded" speed? It will be well to remember that these speeds were calculated by astronomical observation, because the astronomical basis of these calculations will be taken up later in presentation of evidence.

To those who will study up on the subject of Polar Exploration, it will soon become evident that the feature most agreed upon by all North Polar explorers is that the area is oceanic, covered by water, and that it is variously frozen over or partially open, depending on the time of year. One peculiarity which many explorers remark upon, however, is that, paradoxically, the open water exists in greater measure at the nearer reaches of the Pole. In fact, some explorers found it very hot going at times, and were forced to shed their Arctic clothing; there even being one record of an encounter with naked Eskimos. Yet, with all this confirmed oceanic area, we have the contradiction of Admiral Byrd's flight being almost entirely over land, mountains covered with trees, interspersed with lakes and streams.

One of the reports from the Byrd expedition was the sighting of a huge animal with dark fur. Are there such animals, or traces of them, in the Arctic? Beginning in Siberia, along the Lena River,[19] there lie exposed on the soil, and buried within it, the bones and tusks of literally millions of mammoths and mastodons. The consensus of scientific opinion is that these are prehistoric remains, and that the mammoth existed some 20,000 years ago, and was wiped out in the unknown catastrophe we now call the last Ice Age.

In 1799, a fisherman named Schumachoff, living in Tongoose (Siberia), discovered a complete mammoth frozen in a clear block of ice. Hacking it free, he despoiled it of its huge tusks, and left the carcass of fresh meat to be devoured

[19] The easternmost river of the three great rivers of Siberia which flow into the Arctic Ocean.

ch. ends p. 287

by wolves.[20] Later an expedition set out to examine it, and today its skeleton may still be seen in the Museum of Natural History[21] in Petrograd (then St. Petersburg).[22]

Early in the century, approximately 1910, a very scientific meal was served in Petrograd. It consisted of wheat from the Egyptian tombs, preserved foods from Pompeii[23] and Herculaneum,[24] mammoth meat from Siberia, and other interesting and ancient viands.[25] The mammoth meat was fresh, and the mammoth from which it had been taken still had undigested food in its stomach, this undigested food consisting of young shoots of fir and pine and young fir cones. According to the scientists, this mammoth was one of the millions slain instantly in a gigantic catastrophe 20,000 years ago, in a habitat then tropical, in which the vegetation was fern and tropical in nature. Yet, in the stomach of this mammoth is found the sparse food of a sub-Arctic area such as much of Alaska or Northern Canada is today. There is good reason to cast doubt upon the tropic origin of the mammoth, and its sudden demise. And if the demise was not sudden, then the presence of indigested food (not digested even by so much as minutes exposure to stomach acids) in the stomach of the mammoth is unexplainable. True the death must have been sudden, but it was not of tropic locale. If not tropic,

[20] An expanded account of this story can be found in the 1886 book *The Ivory King: A Popular History of the Elephant and its Allies* (Chapter 4: "The Mammoth"; pp.37-39) by American naturalist, conservationist, and author Charles Frederick Holder (1851–1915).

[21] Opened in 1832, the Zoological Museum of the Zoological Institute of the Russian Academy of Sciences is a Russian museum devoted to zoology, and is one of the ten largest nature history museums in the world.

[22] Founded in 1703, and formerly known as Petrograd (1914–1924) and later Leningrad (1924–1991), Saint Petersburg is the second-largest city in Russia.

[23] A city in what is now the municipality of Pompei, near Naples, Italy, which was buried under 13 to 20 feet (4 to 6 meters) of volcanic ash and pumice in the eruption of Mount Vesuvius in 79 AD.

[24] An ancient Roman town located in the modern-day comune of Ercolano, Italy. It was also buried in the eruption of Mount Vesuvius in 79 AD.

[25] Food items.

then the Ice Age onset is not the cause of death. The cause of death, then, is Arctic in nature, and could have occurred any time. But since the Ice Age, there have been no mammoths in the known world. Unless they exist in the mysterious land beyond the Pole, where one of them was actually seen alive by members of the Byrd expedition! Others who dined on mammoth meat were James Oliver Curwood[26] and Gabrielle D'Annunzio,[27] who gave a banquet at the Hotel Carlton in Paris.[28]

We have taken the mammoth as a rather sensational modern evidence of Byrd's mysterious land, but there are many lesser proofs that an unknown originating point exists somewhere in the northern reaches. We will merely list a few, suggesting that the reader, in examining the records of polar explorers for the past two centuries, will find evidences of both fauna and flora impossible to reconcile with the known areas of land mentioned early in this presentation of facts, those areas surrounding the Polar Area on your present-day maps.

The musk-ox, contrary to expectations, migrates **north** in the wintertime. Repeatedly, Arctic explorers have observed bear heading north into an area where there cannot be food for them. Foxes also are found north of the 80th parallel,[29] heading north, obviously well-fed. Without exception, Arctic explorers agree that the further north one goes, the warmer it gets. Invariably, a north wind brings warmer weather. Coniferous trees drift ashore, from out of the north. Butterflies and bees are found in the far north, but never hundreds

---

[26] James Oliver Curwood (1878–1927) was an American action-adventure author and conservationist, in the tradition of Jack London.

[27] Gabriele D'Annunzio, Prince of Montenevoso (1863–1938), was an Italian poet, playwright, orator, journalist, aristocrat, and Royal Italian Army officer during World War I.

[28] The Paris Carlton Hotel, built in 1907, was originally known as the Hotel Windsor. In 1988, it became the headquarters of Air France.

[29] The 80th parallel north is a circle of latitude located 10 degrees south of the North Pole.

ch. ends p. 287

of miles further south; not until Canadian and Alaskan climate areas conducive to such insect life are reached. Unknown varieties of flowers are found. Birds resembling snipe, but unlike any known species of bird, come out of the north, and return there. Hare are plentiful in an area where no vegetation ever grows, but where vegetation appears as drifting debris from the northern open waters. Eskimo tribes, migrating northward, have left unmistakeable traces of their migration in their temporary camps, always advancing northward. Southern Eskimos themselves speak of tribes that live in the far north. The Ross gull,[30] common at Point Barrow, migrates in October toward the north. Only Admiral Byrd's "mystery land" can account for these inexplicable facts and migrations.

The Scandanavian legend of a wonderful land far to the North called "Ultima Thule"[31] (commonly confused today with Greenland) is significant when studied in detail, because of its remarkable resemblance to the kind of land seen by Byrd, and its remarkable far north location. To assume that Ultima Thule is Greenland is to come face to face with the contradiction of the Greenland Ice Cap, which fill the entire Greenland basin to a depth of 10,000 feet. A green, fertile land in this location places itself so deep in antiquity that it postulates an overturn of the Earth, and a new North Pole area (see *National Geographic*'s exploration of the Greenland Ice Cap and its possible significance).[32]

Is Admiral Byrd's land of mystery, center of the great unknown, the same as the Ultima Thule of the Scandinavian legends?

[30] Ross's gull (*Rhodostethia rosea*) is a small gull (the only species in its genus), named after British Royal Navy officer and explorer James Clark Ross (1800–1862).

[31] Thought to be the northernmost region of the habitable world by ancient geographers.

[32] Likely a reference to "Across Greenland's Ice Cap on Skis" by Paul-Émile Victor, published in *National Geographic*'s May 1959 issue.

There are mysteries concerning the Antarctic also. Perhaps the greatest is a highly technical one of biology itself; for on the New Zealand and South American land masses are identical fauna and flora which could not have migrated from one to the other, but rather are believed to have come from a common motherland. That motherland is believed to be the Antarctic Continent. But on a more "popular" level is the case of the sailing vessel *Gladys*, captained by E. B. Hatfield, in 1893. The ship was completely surrounded by icebergs at 43 degrees south and 33 degrees west and finally escaped its entrapment at 40 degrees south and 30 degrees west. At this latitude an iceberg was observed which bore a large quantity of sand and earth, and which revealed a beaten track, a place of refuge formed in a sheltered nook, and the bodies of five dead men who lay on different parts of the berg. Bad weather prevented any attempts at further investigation.[33]

Bear in mind that it is a unanimous consensus of opinion among scientists that the one thing peculiar to the Antarctic is that there are no human tribes living upon it. But this consensus must be wrong, because investigation showed that no vessel was lost in the Antarctic at that time, so that these dead men could not have been shipwrecked sailors. Even today, with Antarctic exploration at its height, the lack of human life on that bleak continent is agreed upon. Could it be that these men who died on that berg came from "that mysterious land beyond the South Pole" discovered by the Byrd expedition? Had they ventured out of their warm, habitable land and lost their way along the ice shelf, finally to be drifted to their deaths at sea on a portion of it, broken away to become an iceberg while they were on it?

[33] An expanded account of the Hatfield story appears in a *Scientific American Supplement*, No. 935 article entitled "Icebergs in the Southern Ocean" by William Allingham, which was published December 2, 1893 (pp. 14947-14948).

ch. ends p. 287

Most recent evidence that there is something strange about the Poles of Earth comes in the launching of Polar orbit satellites. The first six of these rockets launched by the United States from the California coast were full of disappointments—and surprises. The first two, although perfect launchings, seemed to go wrong at the last minute, and although presumed to be in orbit, failed to show up on the first complete pass around the Earth. Technically speaking, they should have gone into orbit but they did not. Something happened, and the location of this something was the Polar area.[34] The next two rockets fired did achieve orbits. This was done by "elevating sights," so to speak, and trying for a higher orbit, with a large degree of eccentricity, that is, a high point of orbit above the poles and a low point of orbit at equatorial areas. It was admitted that this eccentric orbit would produce a short-lived orbit, but it would also give the advantage of readings at widely varied heights above the Earth. Especially interesting was the readings expected above the Poles, because of the discovery of the radiation ring that surrounds the Earth like a huge doughnut, with openings at both Poles.[35] Scientists were very anxious to map this area of low radiation, because it offered a hope of an escape breach for future space travelers who faced almost certain death from radiation while passing through

[34] Discoverer 1 was the first of a series of satellites which were part of the CORONA reconnaissance satellite program. It was launched on February 28, 1959, from Vandenberg Air Force Base in California. It was a prototype of the KH-1 satellite, and the first satellite launched toward the South Pole in an attempt to achieve a polar orbit, but was unsuccessful.
Discoverer 2 was launched on April 13, 1959, and was the first satellite to be stabilized in orbit in all three axes and to be maneuvered on command from Earth. A timing error caused the reentry capsule to land near the island of Spitzbergen rather than Hawaii, and the subsequent recovery operation was unsuccessful. The flight, loss, and search for Discoverer 2 were the inspiration for the book and film *Ice Station Zebra*.

[35] Utilizing instruments containing a Geiger–Müller tube (the sensing element of a Geiger counter) on the 1958 satellites Explorer 1, Explorer 3, and Pioneer 3, during the International Geophysical Year, American space physicist James Van Allen (1914–2006) confirmed the existence of a zone of energetic charged particles now named the Van Allen radiation belt.

the forbidding belt discovered around the equatorial and temperate areas of the Earth.

The next two satellites bore nose cones similar to those in which a future astronaut would be sent into orbit. In each one was a powerful radio transmitter, which was possible because the cone was the size of an automobile, and carried heavy batteries. Also included were powerful lights which could be illuminated at the proper time. The technique of releasing this cone from the satellite was to drop it by a radio-triggered device somewhere above Alaska. Once dropped, the cone lost altitude and proceeded around the Earth for one more revolution on its orbit. Having come over the Pole, it was then low enough (calculated the rocket men) to drop into the atmosphere over Hawaii, where a parachute would lower it slowly to the Earth's surface, and there huge planes awaited, rigged to "fish for" the descending cone, and take it into the plane before it dropped into the ocean and thus retrieve its important contents intact, without damage of crash landing.

On both occasions the following happened: The powerful radio signals were not heard at all. The lights were not seen at all. Radar, with a range at least 500 miles detected absolutely nothing. Each "pick-up" was a complete failure because there was nothing to pick up.

The explanation of the radio failure was advanced as "freezing" of the batteries so that the radio failed to work. No explanation was given for the failure of lights, or of radar detection. That the batteries froze is a strange explanation, considering that similar batteries in other satellites, orbiting for months, and even years, have never frozen. Failure might be admitted in one case, but total failure in both instances bears the aura of improbability.

Each launching was perfect. Orbits finely determined as to exact distance, speed, etc. were achieved, and constantly tracked. Yet, when the final deed is done, and the cone is

ch. ends p. 287

detached successfully according to monitoring devices signaling the detachment, everything goes wrong and the result is complete and inexplicable disappearance of the cone. True, the statement is made that there is only a 1,000 to 1 chance of success, and thus two failures are not unreasonable. But the failures are not to be complete ones. By failure is meant the successful final "pick-up" of the cone by the aircraft. Not complete disappearance! At least radio signals will be received, lights will be seen, radar will spot the descending cone.

Can it be that the reason the descending cone does not come over the Pole on that last low pass is because the Polar Area is mysterious in extent, not in the area calculated by the rocket men, and therefore not taken into consideration? Can it be that the nose cone fell to Earth inside that "land of mystery" discovered by Admiral Byrd? Where else could they have gone? If the Earth at the Poles is as given on today's maps, could four successive "low-level" launchings give the same inexplicable result — unreasonable disappearance?

If there is 1,700 (or more) miles of land extent in **addition** to the area bounded by longitudes and latitudes on a sphere existing in the Arctic, it follows that the recorded disappearances are not inexplicable, but **certain to occur!** Naturally a rocket cone figured to traverse a certain distance (in these cases approximately 33,000 miles) will not land at a predetermined point if the distance to be traveled is greater by 1,700 miles. Our radar will fail to find our cone, and our eyes will see no lights. But why will our radio fail to send its signals to us? Is it because that "land of mystery" is of an "intervening" nature? Radio waves will not go through the Earth, of course. If solid substance intervenes, then we can understand why radio waves do not penetrate it. But what kind of a land configuration can it be that "intervenes" in this way? Why don't we have the "skip and bounce" effect from the stratosphere, which presumably exists over Byrd's "land of mystery" as well as over the lands on the map?

Since the mapped area of the spherical Earth does not allow sufficient room in which to place our two mystery lands, can it be that the Earth is of a different shape, one that allows us to place these lands on that portion of it which does not come under the category of "spherical"? To many readers this will bring a snort and a humorous smile. They will say that we are bringing up the old saw[36] of the Earth not being round. Columbus, they will say, finally proved that to the Earth's peoples, and Magellan actually did sail completely around the Earth by sailing in one direction until he had come back to his starting point.[37] Also, anyone can go out on the night of a Lunar Eclipse, and see the round shadow of the Earth cross the face of the Moon. Seeing is believing, they will say — and just try to get around that!

It is true that seeing is believing, for most people. But the most informed optical scientist will not hold that popular view. He will point out that the human eye, like the telescope, is a lens. And the proved property of any lens is that it tends to make everything look round. No matter what we look at, distance converts the lines of the observed object from straight, or angular, or crooked, to perfectly circular ones. This is a familiar phenomenon to aviators, who know that from the air, no house has a square chimney; they are all round. Any aircraft carrier pilot will tell you that as he brings in his jet at a great height, his carrier looks like a round dot beneath him, and as he descends, it becomes a rectangle again. Anyone who has gone through a railroad tunnel, riding on the rear

---

[36] An "old saw" is an oft-repeated saying, maxim, or proverb.

[37] Ferdinand Magellan (c. 1480–1521) was a Portuguese explorer best known for having planned and led the 1519–1522 Spanish expedition to the East Indies. In 1519, he sailed from Spain, rounding South America through the strait that now bears his name, and reached the Philippines in 1521, where he was killed in a battle, leaving his crew to complete their return voyage and the first circumnavigation of the globe in 1522.

ch. ends p. 287

platform, will testify that the tunnel opening, if square, will gradually grow round as the train proceeds into the tunnel. Optical illusions, they are called. Any camera expert will tell you that the film records distant or extremely small objects as round dots, and that great magnification is necessary to resolve this roundness, and beyond a certain limit (that of the "grain" of the emulsion itself) it is impossible to resolve this roundness.

The scientific fact is that were the Moon actually square, at a distance of 240,000 miles our eye, our telescope, our camera would tell us it is round! No matter what shape it is, we would see it only as round. Actually, we cannot prove that the Moon is round; nor the Earth.

Thus, the arguments for a perfectly round Earth are not based on fact, only on assumption. This assumption is based on a brand of astronomy no longer acceptable to the scientist. Today the nebular theory of formation of planets, suns, even galaxies, is looked upon favorably.[38] The condensation of nebula into stars and planets is accomplished by whirling motion. The whirling motion more often produces the "spindle" shape, round at the "equator," and projecting at the "pole"; or the "doughnut" shape, with flattened poles and holes through the middle. Since the Earth so formed, it may well be that it is either shape. We would not be aware of it by optical evidence, as we have shown.

On the one hand, the "spindle" shape possesses many specific arguments against it, and is the least reasonable. Astronomical bearings taken anywhere on the "spindle" portion would begin to show telltale evidence of the existence of the "spindle" shape. And they would be the reverse of factual

[38] The nebular hypothesis — developed by Immanuel Kant and published in his *Universal Natural History and Theory of the Heavens* (1755), then modified in 1796 by Pierre Laplace — is the most widely accepted model in the field of cosmogony to explain the formation and evolution of the Solar System, which proposes formation occurs when gas and dust orbiting the Sun clump together to form planets.

sightings and bearings taken by Polar explorers. Actually, the bearings taken point to the "doughnut" shape.

Let us go back to Admiral Perry: his astounding rate of travel on his return from the Pole. If he were traveling over the inner lip of a "doughnut" shape, his bearings would indicate a great distance traveled, due to the fore-shortened horizon, and the "expanded" angle used in making his trigonometrical calculations. Actually he would be traveling the same distance each day, and the drop in speed would be entirely compatible with the bearing observations taken with a constantly lengthening horizon.

Rocket scientists have made much of the discovery of the Van Allen Belt, which is a belt of radiation surrounding the Earth. The reader is invited to read about it in *Scientific American*,[39] and especially note the drawings of its shape, which are precisely a vast "doughnut," with the spherical Earth pictured at its center, in the "hole" of the doughnut. What if the Earth is not spherical, but actually doughnut-shaped, exactly as its surrounding Van Allen Belt? Whatever makes the belt thusly shaped, might it not also be responsible for shaping the Earth similarly?

The evidence is extremely strong, and amazingly prolific in scope and extent, that the Earth actually is shaped in this fashion. And if it is hollow, then we no longer need look for the saucers from outer space—but rather from "inner space"! And judging from the evidences, the interior is extremely habitable! Vegetation in abundance is there; animals abound; the "extinct" mammoth still lives! Byrd flew 1,700 miles over the inner edge of the "doughnut hole," and the Navy flew 2300 miles over the opposite inner edge. Both flights went a partial way into the inner Earth. And if this is all true, then no doubt extended flights to 10,000 miles and beyond have been made since 1957 into this hollow Earth, for we have the

[39] Van Allen, J. A. (1959, March). Radiation Belts Around the Earth. *Scientific American* 200(3), pp. 39-47.

ch. ends next p.

planes with the range to do it! If the government knew the significance of the Byrd-Navy flights, it would certainly not neglect to explore further!

Aimé Michel, in his "straight line" theory, proved that most of the "flight patterns" of the flying saucers are on a north-south course, which is exactly what would be true if the origin of the saucers is Polar.[40]

In the opinion of the editors of *Flying Saucers*, this Polar origin of the flying saucers will now have to be factually disproved. It is completely necessary that this be done. More than a simple denial is necessary. Any denial must be accompanied with positive proof. *Flying Saucers* suggests that such proof cannot be provided. And until such proof is provided, *Flying Saucers* takes the stand that all saucer groups should study the matter from the hollow Earth viewpoint, amass all confirmatory evidence available in the last two centuries, and search diligently for any contrary evidence. Now that we have tracked the saucers to the most logical origin (the one we have consistently insisted must exist because of the insurmountable obstacle of interstellar origin which demands factors almost beyond imagination), that the saucers come from our own Earth, it must be proved or disproved, one way or the other.

Why? Because if the interior of the Earth is populated by a highly scientific and advanced race, we must make profitable contact with them; and if they are mighty in their science, which includes the science of war, we must not make enemies of them; and if it is the intent of our governments to regard the interior of the Earth as "virgin territory," and comparable to the "Indian Territory" of North America when the settlers came over to take it away from its rightful

[40] Aimé Michel (1919–1992) was a French science and spirituality author, and UFO specialist, who postulated what he termed as "alignments," straight lines that corresponded to large circles traced and centered on the Earth, and claimed that UFO sightings could be clustered along these grid lines.

owners, it is the right of the people to know that intent, and to express their desire in the matter.

The Flying Saucer has become the most important single fact in history. The answer to the questions raised in this article must be answered. Admiral Byrd has discovered a new and mysterious land, the center of the great unknown, and the most important discovery of all time. We have it from his own lips, from a man whose integrity has always been unimpeachable, and whose mind was one of the most brilliant of modern times.

Let those who wish to call him a liar step forward and prove their claim!

Flying saucers come from this Earth!

FLYING SAUCERS

The Magazine of Space Conquest

FEBRUARY, 1960

WISCO

35ᶜ

U. F. O. PHOTOGRAPHED OVER POLAND

By
Dr. Stanislaw Kowalczewski

★

IS THE EARTH DOUGHNUT SHAPED?

MORE ABOUT POLAR MYSTERY AREA

★

BRAZILIAN TOP SECRET SAUCER REPORT UNVEILED

CONGRESSIONAL U. F. O. INQUIRY DEMANDED

Editorial

Flying Saucers #14, February 1960

The one big flaw in our now "most-talked-of" issue ever, the December 1959 "mystery Polar lands" issue, is the non-existent North Pole flight of Admiral Byrd in 1947. We don't know how many of our readers caught us up on this, and proved that Admiral Byrd did **not** make a flight beyond the North Pole in February of 1947. The fact is, of course, and we admit it freely, that the February 1947 flight was a **South** Pole flight, just as was the 1957 flight. The only flight Admiral Byrd made over the North Pole was in 1926, and in it, he merely flew to the Pole, **circled** it twice, then returned to Base.

Almost unanimously, our readers who detected this flaw asked us **how** it could happen that we could make such a serious error, after our years of research into this thing, and particularly since we said we wanted to be sure of our facts before we presented our "Earth is not round" theory. They pointed out that we had seriously impaired the validity of our theory: if the North Pole is perfectly normal, then our theory about the South Pole might be damaged.

In this issue, we are writing the Editorial last. In the back of the magazine you will find many letters, the first of a deluge we received, with many comments. It should be known that the reason they are printed there is the matter of publication deadline: they were the first available, and already late with the issue, we used them. Some of the letters now at hand are much more interesting, but they will have to wait until next issue. Suffice it to say that they will be published, plus many more we expect to get. They will prove intensely interesting reading. So, if there exists a paradox between this editorial and its comments, and the letter section and the editorial answers to some of the letters, this is the reason.

Now as to the **how** of our "serious error." Some time ago, we made a remark that there was a systematic effort being made to render the whole flying saucer story ridiculous. We promised to make this a particular point of attack in the future, and name names, present facts. That time hasn't come yet. We can only say that we haven't changed our mind about this effort, and now, with the publication of the December issue of *Flying Saucers*, and our claims in it, we are more certain than ever. In short, what we want to say is that our statement, that Byrd made a North Pole flight in 1947 to 1,700 miles **beyond** it, was a fishing expedition, for which we hope our readers will pardon us.

Not only has it not weakened our theory, it has strengthened it immeasurably, and also strengthened the theory that a systematic effort is being made to render flying saucers a subject of ridicule. In short, there were two alternatives—either Byrd made a **secret** flight over the North Pole in 1947, which **never** hit the newspapers, or a deliberate effort was being made to build an edifice that could be toppled **if and when the truth came out about the South Pole**! There was only one way for this editor to discover if he had, somehow, missed the **big** story, the actual fact of a 1947 Byrd Polar flight as described, and by publication, he could ferret out the missed story. The whole thing was what you might term a "calculated risk." As it actually turned out, Byrd did **not** make a North Polar flight 1,700 miles beyond the Pole—but he **did** make a South Polar flight 1,700 miles beyond the **South** Pole! We had hoped that more of our readers would notice that we had, in our research, not discovered the 1947 South Polar flight, which would be strange indeed, because it sticks out like a sore thumb all over the newspaper, magazine, and book world! Reams have been written of that flight. Our alibi, if we wished to use one, could have been that it was simply a typographical error, and we meant South Pole all the time, and not North. But then somebody would have popped up and said—but this is ridiculous, because of all your specific mention of the areas around the **North** Pole. No, we aren't presenting any alibi—what we are presenting is possibly the missing fact we were searching for all along: that to this day, no attempt has been made to fly **beyond** the North Pole, but that every trip to the Pole has followed the standard navigational procedure which we pointed out must invariably turn you away from any land "beyond" and bring you back on a **southerly** course which will bring you inevitably to a known land. Byrd, in 1926, flew to the North Pole, circled it twice (it took about five minutes to fly these two circles) and then headed back to his starting point!

When we read of Admiral Byrd's amazing 1947 **North** Pole flight, 1,700 miles **beyond** the Pole, and compared its startling similarity to the actual 1947 **South** Pole flight, which went the same distance beyond the South Pole, we found ourselves with an obvious paradox, yet not an impossible one. There were two weeks **possible** difference in time of both flights, and Byrd **could** have made them both. The question was, **did he**? We had absolutely no way of knowing. We reasoned thusly: If by chance Byrd did make such a North Pole flight in February of 1947, and it was actually kept secret, but in some obscure publication it "leaked" out, but was covered up so successfully that we had been unable to discover the "leak," we might find one of our readers sending us the material in question, possibly dug from yellowing newspapers in

ch. ends p. 301

his attic. No such confirmation was forthcoming, but much information correcting our error was.

Where did we get our information on Byrd's non-existent North Pole flight and 1,700 mile penetration beyond it in 1947? Yes, it **was** published, and in a very strange book. We feel that now we can reveal the book, and its information, and ask our readers but one question—isn't this the **perfect** book to make a laughing stock out of anyone who happens to come up with our "mystery land at the Poles" theory?

We doubt if this book has enjoyed a large sale. We doubt, in fact, if it has achieved even a slight sale. Our two copies came to us as "review copies" from the publisher. We have not seen it publicized anywhere[1]—we have not heard of any of our readers buying it. Certainly none of them wrote us and mentioned it in connection with our story in the December *Flying Saucers*. That its distribution is minor indeed is apparent. However, perhaps our mention of it here will stimulate sales. In fact, it should sell the edition out! For which the author, if he is on the level, should thank us profusely. The title of this strange book is *Worlds Beyond the Poles* ($3.50), its author F. Amadeo Giannini. It was published by Vantage Press, Inc.,[2] 120 W. 31st Street, New York, N.Y., on July 6, 1959. In it, not only is the 1947 **North** Pole flight of 1,700 miles **beyond** the Pole in a northerly direction completely described, but it is reiterated over and over again, with complete quotes from newspapers, radio, etc.

Because of this positivism, your editor could not conceive that either Giannini or Vantage Press could have made so disturbing a "typographical error" because of the very facts

[1] Granted, the world wasn't as connected in 1960 as it is today, so Gianinni, then, wasn't just a Google search away, but based on our research for this edition, Giannini's book was publicized as well as could be expected for an alt-theory book published by a small press in 1959. The many newspaper clippings and ancillary articles that we have included prove that.

[2] Vantage Press, Inc. was a New York City-based publishing company, primarily known as a vanity press, operating from 1949 to 2012, and gained notoriety for its "cooperative" publishing model, where authors paid for the costs of publication, rather than the publisher investing in the project.

we mentioned before which would render the same claim on our part invalid, namely the direct reference to North Polar adjacent lands. There were two possibilities—that Giannini was deliberately falsifying, or it was true that he had access to information your editor did not, and in fact, could not verify, no matter how intensely he searched. There was only one thing to do—to pass this on as my own information, so that one would suspect the existence of the Giannini book, and discover, if I could, if anybody **else** than Giannini had possession of actual information regarding the contradictory flight. Since nothing has turned up (and it should have, if true, because our readers turned every library in the country upside down in an effort to either confirm our claims, or to disprove them), we can only conclude the first alternative: that Giannini deliberately falsified. The big question is **why**?

Briefly, for our readers' information, Giannini claims the Earth is just a part of a **continuous land area** which comprises the entire universe, and that the access to such seemingly disassociated worlds as Mars, the Moon, other planets is via the Poles, which are continuous **terra firma** extending upward and outward into "space" so that the only sensible way to go to the moon is not by rocket, as the Russians did, but by motor car (or tractor if the going is rough) into the sky on the spindle-shaped Poles of the earth.

It is rather hard to describe Giannini's concept of "space" and "land" (known and unknown), and even his diagrams intended to illustrate the stratosphere, the true nature of stars and planets, and that up is not really up at all, but along some strange parallelity of opposite and facing land masses somehow connected at the converging end of "parallelity" via a mysteriously shaped Polar access, only add to the confusion. In fact, it is not visualizable, and Giannini admits that only physically can it be visualized.

Yet, throughout the book there exists the whole gamut of strange facts which we ourselves had been aware of for years, **all** carefully mustered to support a theory doomed

ch. ends p. 301

by every process of logic to be forever incomprehensible. No one, reading the book, will seriously hold the "land continuity" theory he expresses for even a moment. In fact, if he reads the entire book, he will once and for all, faced with **legitimate** points in the book, refuse to consider them for any **other** purpose because of the illegitimacy that has "rubbed off" on them by association with the weird "continuity" theory Giannini apparently **seriously** puts forth.

Perhaps the reader can imagine the emotions of this editor, after twenty years of careful assembly of facts preparatory to launching a theory of the lack of true concept of the shape of our Earth, when he found many of the impressive items in his arsenal **firing resounding salvoes**[3] **of blanks** in support of the utterly ridiculous. Now, to use the same items in support of what amounts to an exactly diametric theory: i.e., that the Earth is strangely shaped at the poles, either in a dish-shape depression of huge dimensions, or a hole all the way through, with the resultant Earth distinctly doughnut-shaped, seems almost folly. Ho, ho, ho—first Giannini, and now Palmer! One on a spindle to Mars, the other in a hole to nowhere!

Now, we think, the strange book written by Giannini has offered the one possibility by which it can definitely be proved that the Earth is shaped strangely at the North Pole, as we believe it to be at the South Pole, not necessarily with a hole all the way through, but like a doughnut which swelled so much in cooking that the hole is only a deep depression at each end, or like a gigantic auto tire mounted on a solid hub with recessed hub caps. The fact is, now that it seems sure that Giannini falsified the Byrd North Pole flight, no human being has ever flown directly over the **North** Pole, and continued straight on, as has been done at the South Pole. Your editor thinks it **should** be done, and done immediately. We have the planes to do it. There are dozens of volunteers

---

[3] The simultaneous discharge of weapons or artillery.

among *Flying Saucers*' readers who would be delighted to accompany the brave men of the U.S. Air Force in a "round the world" bomber on a trip of this kind. Your editor wants to know **for sure**, whether such a flight would wind up in any of the countries surrounding the North Pole, necessarily **exactly** opposite the starting point. Navigation is not to be made by compass, or by triangulation on existing maps, but solely by gyrocompass[4] on an undeviatingly straight course from the moment of take-off to the moment of landing. And not only a gyro in a horizontal plane, but one in a vertical plane also. There must be a positive forward motion which cannot be disputed.

Everyone knows that a horizontal gyrocompass, such as used now, causes a plane continually to gain in elevation as the earth curves away below it as it progresses. These gyrocompasses are either fixed to compensate for this tendency toward flying even higher, or the pilot himself continually adjusts to it by keeping "on" his altimeter at all times. That is the reason for the vertical gyrocompass. Only with such a compass can an undeviating course, neither to right or left, be maintained.

Because of the implications which our Polar Land theory has inherent in it your editor would be **greatly relieved** to have the whole thing definitely disproved, and the conventional spherical Earth established as a positive fact. But simply to tell us to "go back and sit down and forget about it" is unacceptable. Perhaps the Air Force will look upon a child who insists that he dig a hole fifty feet deep to prove to him that the pebbles at that depth are not really diamonds set in golden crowns. The father knows they are the same pebbles as at the surface, and the prospect of digging so fruitlessly is exasperating. But the Air Force is not like the father; they are not averse to flying millions of miles, spending millions of

[4] A nonmagnetic compass that uses a continuously motor-driven gyroscope, whose axis is parallel to the earth's axis of rotation, to maintain the direction of true north.

ch. ends p. 301

dollars of the taxpayers' money to chase down **non-existent** (so they insist) flying saucers; so why should they object to a simple $100,000 flight across the Pole on a straight line from one known civilized area to another? When you consider that "navigationally" it hasn't **really** ever been done in the case of the North Pole, at least, it would be interesting to do it, and log one more splendid achievement by the Air Force.

Many of our readers stated that even commercial flights continually cross the Pole and fly to the opposite side of the Earth. This is just not true. Even though the airline officials themselves, when asked, might say they do, it is not **literally** true. They **do** make the navigational maneuvers which automatically eliminate a flight **beyond** the Pole in a straight line, in every case. Ask the **pilots** of these Polar flights. And when you come to pin it right down, name **one** trans-polar flight on which you can buy a ticket **today**. Not just a flight from New York to Gander, Newfoundland,[5] and on out over the Atlantic to Ireland, or any other flight you can find on the travel schedules of any airline in the world—but a flight which **actually crosses** the North Pole.

Finally, will someone settle the argument about **where** the two **magnetic** poles actually are? We have had dozens of letters, all stating in equally **positive** terms, dozens of different locations for the magnetic poles. It seems that some experts even say it wanders about from day to day, as much as hundreds of miles! **Is** the magnetic Pole a **point**, geographically, or is it a will-o'-the-wisp?[6] Is it possible for one man to stand "at the pole and be a thousand miles from another man, also standing at the pole?" Is the magnetic pole actually a gigantic ring around the upper lip of the "doughnut?" Is it more or less a "center of magnetism" such as there is a

[5] Gander is a town located in the northeastern part of the island of Newfoundland in the Canadian province of Newfoundland and Labrador, and is the site of Gander International Airport, once an important refueling stop for transatlantic aircraft.

[6] In this context, something that is elusive or impossible to find.

"center of gravitation"? Just where is the center of gravitation of a doughnut? The "center of gravitation" of **two** bodies (say Mars and Earth) **is somewhere between them**. Actually, gravitation, magnetism, etc., are rather evanescent[7] things, and when we measure them and locate them, we are possibly using measuring sticks lacking in the necessary "dimensions" to give us a true answer. They say the South Magnetic Pole is 2,300 miles away from the South Geographic Pole. But nobody says **why**! If the Earth is perfectly spherical, and its electromagnetic field is also spherical, the positive and negative poles must necessarily be exactly opposing, and diametrically opposite. If one pole is located on a precise point on the sphere, the opposite pole must be **exactly** opposite. If magnetism depends on something else than the total mass of the planet, but rather on locations of major iron ore deposits, then we can have poles all over the place.

As a rather random point, let us assure our readers that we did not say the Moon was square! We only said that if it **was**, you couldn't prove it. If the moon is anything but round, as we see it, we prefer to think of it as doughnut-shaped, but not square. Our purpose in our analogy was only to state that appearances can be deceiving, and actually are. It comes to the reader who visualize himself as standing five hundred miles high, and thus obviously able to "see" the hole at the Pole, if he is standing at its edge. When picturing this doughnut-shaped earth, and its "hole" in the doughnut, picture yourself a normal six-feet or less in height, and try to get a truer perspective. You could no more see the "hole" than you can the Earth's curvature from your six-foot height!

The two men who recently photographed Venus from a balloon at above-stratosphere heights made two interesting observations: they said that there was water on Venus, but they would not say how much; and that it was totally dark

[7] Quickly vanishing or fading away.

ch. ends p. 301

in space, at their height.[8] No one seems to have noticed that although they were in brilliant sunlight, yet there was darkness so that they could not see without manufactured light of their own. Down below, they could see where the stratosphere began, or where **light** begins, and the surface of the Earth. It would be the same at the Poles, as we progressed inside the Earth (if it is indeed hollow). We would see **upward** only to a height of 10 miles (the stratosphere). Looking beyond that we find nothing but darkness. Light is manufactured in that first 10 miles of the Earth's atmosphere. Considering this mysterious fact, we can picture the inside of our Earth filled with daylight due to the action of the atmosphere, and fluctuating electromagnetic **field**, which causes it to **fluoresce**,[9] and not to the presence of the sun at all! Perhaps the sun's light is only a focal point of fluorescence caused by the interaction of the combined electromagnetic fields of both the sun and the Earth?

More and more we are learning that the exploration of space is suggesting that our "facts" of yesterday are only theory based on an insufficiency of evidence; and as we learn more, we are forced to revise our theories.

This editor thinks that what we are learning at the Poles is forcing us to revise still more "theories" about this old Earth of ours. So important is the South Pole area (at least) that 12 nations, including Russia, with one accord, without dissent, agree to the most amazing peace pact ever conceived. The Antarctic is **never** to be used for military purposes, or as a rocket launching site, but solely for scientific (mutual) endeavor.

---

[8] Responding to a challenge by fellow professor James Van Allen (see p. 280, footnote #35), German-American astrophysicist Martin Schwarzschild (1912–1997) conceived balloon-borne astronomical telescopes known as Stratoscopes. Project Stratoscope I possessed a 12-inch (30.48 cm) mirror and was first flown in 1957. A small secondary mirror focused the image from the primary into a 35mm movie camera, which captured the images on film.

[9] To shine or glow brightly due to fluorescence.

Just the other day we picked up a newspaper, and found a well-known columnist remarking that there must be something peculiar in the Antarctic that causes this mysterious instantaneous and mutual accord among nations whose entire instincts should be to acquire personal supremacy in the South Pole area because of its tremendous personal significance, which is 90% significant to the Russians and only 10% significant to any other nation! Only Russia could gain advantage in rocket distances to enemy lands by owning the South Pole! For any other nation to go there to launch rockets at Russia is only to increase the distance between their rockets and Russia. Then, if this is true, why did Russia, with not one word of dissent, not one single remonstrance, no arguments whatever, agree to this peace pact, this non-use of the Antarctic for military and specifically military rocket usage?

There has **got** to be a good reason. And it is becoming increasingly obvious even to the average newspaper columnist. All over the Earth there must be wrinkled brows as logical men ask themselves for the answer to the illogic of the Russian action.

With Admiral Byrd, we repeat: "that land beyond the pole, the center of the great unknown!" It isn't the Pole that is the center of the great unknown, but the land beyond it.

In closing this editorial, we want to dwell for a moment on the fantastic bad luck that has dogged us since our printing of the December issue. First, we sent out 54,000 mailing pieces, based on the availability of our over-run of the December *Flying Saucers*, and offered to begin the new subscriptions with that all-important December issue. Also, we held approximately 2,000 current subscribers' envelopes at our office, for personal transmission of their copies of this issue because we wanted to remind them to renew. For two weeks we waited for the truck which delivers our over-run to us

ch. ends next p.

from Sandusky, Ohio,[10] where the magazine is printed. We weren't worried, because the truck service is notoriously slow. But when the truck came, there were no magazines. A phone call to the printer revealed that not only did he not have thousand of copies at the plant, he could find no waybill[11] proving shipment to us. Ordinarily, since we paid for the magazines, we would ask that he return the plates to the press and run off the copies due us. But strangely, this time the plates are not available, are so badly damaged by some careless treatment that no re-run can be made. Where are those thousands of magazines, which were certainly printed, because they are printed along with our other magazine, *Search*,[12] and in the same quantity, and *Search* is entirely accounted for? Why is there no waybill? Even if the waybill was lost, the magazines would turn up somewhere; someone would refuse them if shipped to them by mistake.

Thus, we now have nearly 5,000 irate subscribers who have not gotten the magazine they paid for. They will probably not get it until **after** they get **this** issue, because our only recourse was to ask our distributors to return full copies to us for credit instead of just returning the covers (all at our shipping expense). Another distributor, receiving 750 copies, turns up missing, leaves no address, yet the 750 magazines cannot be located. If not deliverable, they are returned to us, since postage for such return is guaranteed and return requested. Yet, no trace of the magazines. They could not have been delivered to a distributor who had "left town" leaving no address. They did not come back.

Trouble, trouble, trouble. All we can say is for our old and

---

[10] Situated on the southern shore of Lake Erie, Sandusky is located roughly midway between Toledo and Cleveland, and was known as a center of paper-making.

[11] A document given by a carrier to a shipper acknowledging receipt of goods being shipped and specifying terms of delivery.

[12] Founded by Palmer as *Mystic* in November 1953 and renamed *Search* in October 1956, it eventually merged with *Flying Saucers* in 1976, and ceased publication in 1977.

new subscribers to be patient; we will eventually get your copy to you. If not we will reprint the whole thing no matter what it costs us!

Daughter Jennifer[13] broke her leg too, cutting available time to devote to editing and publishing to less than half normal, and we never have any time to spare anyway! Thousands of dollars in hospital and doctor bills. Losses on the magazine.

We could blame it all on Shaver's Deros,[14] but we won't. No streak of bad luck can stop our interest in the mystery lands at the Poles! Future issues of *Flying Saucers* may be erratic, due to circumstances beyond our control, but they'll be printed! And they'll contain an amazing array of questions and answers that you won't want to miss. All we ask is that every one of you who is truly interested do what you can to dig up more and more facts and transmit them to us for publication. Many researchers are needed. Every clue is valuable.

And we've got a real surprise coming up. Not all our luck has been bad—one bit has been fantastically fortunate! Almost beyond hoping for. We firmly believe that 1960 will be the most memorable year in our publishing career, devoted as it has to a lifetime of searching for truth.—Rap.[15]

[13] While we Heathens usually provide biographical information for anyone mentioned in our editions, our research revealed that Jennifer is still living and we're choosing to respect her privacy.

[14] A reference to the Shaver Mystery, of which Palmer played a significant role in its proliferation, which began with a letter that Richard Sharpe Shaver (1907–1975) wrote to *Amazing Stories* in 1943, wherein he claimed to have discovered an ancient language called "Mantong," and advanced prehistoric races who had built cavern cities inside the Earth. Some of their offspring were noble and human "Teros," while most degenerated into a population of mentally impaired sadists known as "Deros."

[15] In editorial writing from the mid-20th century, especially in magazines or newsletters with a casual or conversational tone, the word "Rap" was sometimes used to mean a chat, informal talk, or commentary. It comes from the older slang usage of "rap" as a term for talking or discussing something earnestly or informally, so when an editor signed off with "Rap," they were signaling that what preceded was their candid opinion or editorial voice—like saying, "That's my piece" or "Just my two cents."

FLYING SAUCERS

The Magazine of Space Conquest

WISCO

JUNE, 1960

35¢

How Atlas' Camera Lies To Us!

Pictures purporting to show curvature of earth prove illusory!

RUSSIAN HEROES—OR SACRIFICIAL VICTIMS?

THE MYSTERY SATELLITE SNAFU ★ SAUCERS SERIOUS BUSINESS—Air Force

---

UFOs—FROM THE CRITIC'S CORNER

Editorial

Flying Saucers #15, June 1960

There are several points which it is necessary to clear up in the most prominent position possible, and the beginning of the editorial is about as prominent as we can get it.

1. This is the issue of *Flying Saucers* which follows the February 1960 issue, and since February was number 14, this is number 15. It is dated June, 1960, which means the April 1960 issue **did not** get printed! So, if you feel inclined to write and inform us you did not receive your April issue, please don't. We've got enough unanswered mail as it is!

2. With the publication of our December 1959 issue, troubles began to pile up, but we want to make it perfectly clear, we were **not** visited by any men in black, nor were we silenced in any respect by any authority, governmental or otherwise. It is true that thousands of our magazines were mysteriously missing in transit, but the shortage was remedied by calling in unsold copies from newsstands. Thus it may be that you had trouble finding a copy at the newsstand at which you regularly buy your copy, but if so, we can fill your order from our stock.

3. We don't intend to quit publication! If by any chance, *Flying Saucers* ceases publication without a good reason given by your editor or his family, it will mean simply that everything we've said in the pages of *Flying Saucers* is true, and that some powerful agency is attempting to keep these things secret. Is that plain enough? There are only two things that can make us voluntarily give up *Flying Saucers'* publication, and they are extreme old age, or a million dollars (in which latter case we will put out a magazine that will really bug your eyes!)

4. Several rather prominent flying saucer personalities have written us in the following vein: "You sure ruined yourself with that fake North Pole trip of Admiral Byrd. Too bad — you might have gotten somewhere if you hadn't gone off the deep end!" To this we have only one reply: "Admiral Byrd did make a polar flight in 1947 exactly as described with one exception — he made it to the South Pole, not the North. And as for getting somewhere, we really have, now! And how!"

With the publication of this June issue of *Flying Saucers*, we have on hand such a tremendous mass of material that we can guarantee one thing — the next ten or twenty issues are going to be the most exciting publishing adventure in our long career! Anybody who is at all interested in flying saucers (to cover just one phase of the whole gigantic phenomenon) should not miss a future issue no matter what he thinks as of now. This isn't a sales pitch, because we're convinced we'll be working for nothing the rest of our lives, and we just don't care; what we've got to say has got to be said.

At the same time, with publication of this issue of *Flying Saucers*, we know full well that the content is nowhere near what we wish it was, for a whole host of vexing reasons, but we do think it is a very fine issue, and that it is the forerunner of much finer to come. It is true that this issue contains very little that actually can be said to follow up our Polar Theory. If we had a large staff of writers to whom we could pass out the material being received, and assign them particular aspects of it to whip into presentable shape, we could fill the magazine permanently with it. But your editor is one man, with a whole host of things to do, and he works on the theory that accomplishment is a matter of putting one foot in front of another, one at a time, and that is precisely what we are doing. If you stick with us long enough, our destination will become visible and obvious, and totally interesting.

But, with the space left us in this issue, let's just touch upon a few interesting and intriguing points. The first has to do with the interesting theory advanced by several of our readers, (and backed up with scads of mathematical calculations too!), that the North and South Magnetic Poles may be focal **points** of magnetism on a giant **circle** which progress at a prescribed pace just as the equinoxes, the constellations through the zodiac, the march of the stars about their spatial pathways. This progression along this circular pathway is even placed at about 12 miles per year, although we personally don't state that as any correct figure at this time — it could be slower or faster. The point is that the Magnetic Poles move steadily in a kind of earth-surface orbit around the Geographic Poles. At this time we will give only one bit of information, from among the dozens that have come to us, that points to merit in this theory. We would like to quote from the *Naval Aviation News*,[1] page 18, January 1960 issue:

"The 'Lost Continent of Antarctica' came up with two more [more? —Editor] debits on November 1 when a Navy **Skytrain**[2] reported the disappearance of two tall mountains.

[1] The U.S. Navy's oldest periodical, published quarterly, tracing its origin back to the Chief of Naval Operations' Weekly Bulletins that began in 1917.
[2] The Douglas C-47 Skytrain is a military transport aircraft. It was used extensively by the Allies during World War II for troop transport, cargo, paratroopers, towing gliders, and military cargo parachute drops.

ch. ends p. 317

The discovery was made during a routine aerial reconnaissance by a plane of VX-6.[3] It flew over the zig-zag route to be followed in the next three and one half months by a nine man traverse party[4] in Marie Byrd Land.[5] One of the mountains, Mt. Vinson,[6] was charted during **Operation Highjump** in 1946-47.[7] Mr. John Pirrit,[8] glaciologist and leader of this year's traverse party, believe the 20,013-foot mountain non-existent as a result of observations made last year. Aboard the **Skytrain** he saw an aerial confirmation of his suspicions. Thirty miles further, the 15,000-foot Mt. Nimitz[9] also failed to materialize as the Skytrain flew over its charted position. It is understandable how Mt. Vinson got on the chart, Mr. Pirrit said. By flying over the Executive Committee Range,[10] an aberrated image of Mt. Sidley[11] is

[3] Air Development Squadron Six (VX-6 or AIRDEVRON SIX, commonly referred to by its nickname, "puckered penguins") was a United States Navy Air Development Squadron based at McMurdo Station, Antarctica, whose mission was to conduct operations in support of Operation Deep Freeze, the operational component of the United States Antarctic Program.

[4] A group organized to conduct a traverse survey (a surveying method used to establish control points and map an area).

[5] An unclaimed region of Antarctica, and the largest unclaimed territory on Earth, named after Rear Admiral Richard E. Byrd's wife.

[6] The name was reused for another Mount Vinson or Vinson Massif, elevation 16,050 feet (4,892 m), located within the Sentinel Range of the Ellsworth Mountains, and named after Carl G. Vinson (1883–1981), United States congressman from the state of Georgia, for his support for Antarctic exploration.

[7] Organized by Rear Admiral Byrd, Operation Highjump, officially titled The United States Navy Antarctic Developments Program, 1946–1947, (also called Task Force 68), was a United States Navy operation to establish the Antarctic research base Little America IV, and included 4,700 men, 70 ships, and 33 aircraft.

[8] T. John Pirrit (1924–1962) was a Scottish glaciologist known for his work in Antarctica, particularly his involvement in traverses across Marie Byrd Land and his role at Byrd Station. The Pirrit Hills in Antarctica, an area of peaks and nunataks, were named after him. Also, his death seems sus.

[9] The name was reused for what is now known as Nimitz Glacier, located in the Ellsworth Mountains, and named after Fleet Admiral Chester Nimitz (1885–1966) in recognition of his role in Operation Highjump.

[10] A range consisting of five major volcanoes in Marie Byrd Land.

[11] Mount Sidley is the highest dormant volcano in Antarctica and the highest of the five volcanoes that comprise the Executive Committee Range.

seen. Mt. Sidley is about 180 miles from Byrd Station.[12] Deduction therefore was that Mt. Vinson must have been a mirage."

In this report we are asked to believe that in 1946-47 Operation **Highjump** charted two mountains which they named after Admirals Vinson[13] and Nimitz, established the height of Mt. Vinson to be precisely 20,013 feet, and the height of Mt. Nimitz to be 15,000 feet (neither to be considered small mountains by any stretch of the imagination), and 12 years later these mountains are nowhere to be found, and in fact are explained away as "aberrated mirages." Not just mirages, but crazy ones! If your editor were a member of the original charters of these mountains in 1947, we would demand satisfaction for this insult. We cannot believe that the scientists of Operation **Highjump**, whose accomplishments can be reviewed in detail in the *National Geographic* (to give the easiest place for reference), were so crassly incompetent. We prefer to believe that Mts. Vinson and Nimitz were exactly where the 1947 cartographers placed them, and that they are still there. We predict that they will be rediscovered, and will be said to be several hundred miles away from their originally charted location. (But everybody knows mountains don't move!) It may be that they will not be recognized then, but certainly their original discoverers took pictures of them. If anyone can provide us with pictures of these two mountains, and a picture of Mt. Sidley, we will be glad to publish them all, to prove that they are three different mountains, and that two of them are not "aberrated images" of the third!

There was only one way to chart these mountains, and that was by use of the magnetic compass and with an assist

[12] Former research station established during the International Geophysical Year by U.S. Navy Seabees during Operation Deep Freeze II in West Antarctica. Built in 1957, it was a year-round base until 1972, then seasonal until 2005, when it was abandoned.

[13] Vinson was a politician, not an admiral.

ch. ends p. 317

from the sun and stars. Since the sun and stars are totally unreliable in these far southern areas as a means of precise location, it would be the major role of the compass to give direction, to make the role of the sun in giving latitude and longitude at all positive. Thus, if the South Magnetic Pole travels about on its circular orbit at the rate of anywhere from 12 to 18 miles per year, it is certainly to be expected that an expedition twelve years later, charting its way southward, will wind up more than a hundred miles away from the mountains in question, while believing themselves to be unquestionably in the same spot!

It is this mysterious factor that has lead to so many arguments among polar explorers. Lands discovered and charted have later been impossible to find within thousands of miles of their original supposed locations. Over the hundreds of years of North Polar exploration, succeeding explorers have often marveled at the evident fact that their predecessors could have been so wrong in their mapping efforts.

As an interesting aside, the *Naval Aviation News*, same page, tells of two types of insects which are regularly collected on three-hour bug-runs by sleeve-nets thrust from the doors of a single engine UC-1 Otter.[14] One of the bugs is a small wingless fly and the other is called a springtail.[15] The theory is advanced that these insects are blown into the deep freeze of Antarctica by prevailing winds, for certainly they cannot live in this area of temperatures down to 100 degrees below zero and lower. It is interesting to note that it is a wingless insect that is picked up in some warmer clime

[14] A single-engined, high-wing, propeller-driven, short take-off and landing (STOL) aircraft developed by De Havilland Aircraft of Canada Limited (DHC), the Otter had a reputation for ruggedness and the ability to lift a large amount of cargo from makeshift airfields. UC-1s were crafted specifically for the U.S. Navy.

[15] Any of numerous small wingless hexapods of the class Collembola, having the ability to jump by means of a forked appendage on the abdomen that acts as a spring; found in soil rich in organic debris or on the surface of snow or water.

and deposited here in Antarctica. This is a "ground-scooping" wind, we are forced to hazard, and we wonder why it is that hundreds of other species of tiny wingless (and why not winged also?) insects are not likewise scooped up by this prevailing wind and brought hither? Also, if this is the answer, why waste time in regular three-hour "bug-runs" to collect the little beasties? A great deal like a camel straining at a gnat! To what purpose? Once you've got one bucketful, do you try for a hundred? Perhaps springtail soup is delicious!

On March 4, 1960, Charles Morris, Dubuque, Iowa, airplane instructor saw flying saucers. It was at sundown, and he saw three silver saucer-shaped objects whirring along at about 200 miles per hour, about 20,000 feet up. Morris can prove he saw them because he got 19 feet of film with his movie camera. That is, he could have proved it, if he had kept the film. But he gave it to the "federal government" (Air Force?) for processing and study (and safe keeping, we presume!). Too bad, Morris — you might as well get used to it — you didn't see a blame thing! What pictures, Morris? Let us know when you get 'em back, will you? And if you do, prove they are the same pictures, and that they are intact. Whoops, there goes another rubber-tree plant![16]

Remember Edward Ruppelt, whose book, *The Report on Unidentified Flying Objects*[17] sold 50,000 copies, and which contained so much inaccurate information about yours truly? Well, the book has been revised and reissued, and now, whereas the first edition said "they is," now he says "they ain't."[18] "Flying Saucers were (note the past tense) the

[16] According to an official report: "The 8 mm kodachrome film, which he exposed in late afternoon, failed to show the UFOs."

[17] *The Report on Unidentified Flying Objects* is a 1956 book by then-retired Air Force UFO investigator Edward J. Ruppelt (1923–1960), detailing his experience running Project Blue Book, the code name for the systematic study of UFOs by the U.S. Air Force from 1952 to 1969.

[18] In 1960, Ruppelt published a revised edition wherein he reversed his earlier openness to the UFO question and stated he was "positive" that UFOs did not exist. He also never mentions Palmer anywhere in the book.

ch. ends p. 317

illusions of people who didn't understand what they really saw: weather balloons, passing aircraft, stars, etc." The book is to be translated into French, German, and Portuguese, and published abroad. As a publisher, your editor wonders why? Not to make money, certainly, because it can't possibly! Mr. Ruppelt is a resident, by the way, of Long Beach, where he draws the long bow — which we've gone into before regarding the things he said about your editor, and by golly, there is one place we can stand on firmly! We were **there**, Ruppelt, and you **weren't**! So much for your new book to brainwash the peasants (that's us).

In 1911 Explorer Robert F. Scott tried to reach the South Pole on foot.[19] At one point, apparently, he stopped to build a hut,[20] because on February 1, 1960, it was reported by Professor Robert L. Nichols,[21] head of the Tufts College–National Science Foundation Expedition, that his five-man party had found the hut, and inside it found books, shoes, spice boxes, and a tobacco tin used by the English explorer, who lost his life in 1912.[22] Thus far, we haven't been able to reach Professor Nichols to ask him what the hut was constructed of. If of wood, where did Scott get the wood? And if of snow blocks, how did Nichols find it? Even Byrd's camp, which everyone knew how to locate, was buried deep beneath the surface, in perpetual ice, and they had to tunnel down to it. What chance did Nichols have to find this hut? But he did, apparently, and it is called a "hut." Our curiosity is aroused! If he used native wood to build it, we are not being told all the details. Was the hut in a warm area where trees grew,

[19] Robert Falcon Scott (1868–1912) was a British Royal Navy officer and explorer who led two expeditions to the Antarctic regions: the Discovery expedition of 1901–04 and the Terra Nova expedition of 1910–13.

[20] Scott's Hut (at right) was built in 1911 by the British Antarctic Expedition of 1910–1913 (also known as the Terra Nova Expedition) on the north shore of Cape Evans on Ross Island.

[21] Robert L. Nichols (1904–1995) was a professor and pioneering American geologist known for his research in volcanology and polar geology.

[22] Many of these items are still found inside the hut even today!

Scott's Hut in 1913

CAPE EVANS

THIS HUT WAS BUILT BY MEMBERS OF THE
BRITISH ANTARCTIC EXPEDITION 1910-1913
UNDER CAPTAIN R.F. SCOTT
IN JANUARY 1911

THE HUT WAS RESTORED IN 1960 AS NEARLY AS POSSIBLE
TO THE CONDITION IT WAS IN WHEN OCCUPIED.

VISITORS ARE ASKED TO REMEMBER THAT THIS BUILDING IS A
HISTORIC SHRINE. ITS CONTENTS ARE IRREPLACEABLE.
PLEASE DO NOT INTERFERE IN ANY WAY WITH ARTICLES IN
THIS HUT.

NOTHING WHATEVER MUST BE REMOVED.

ABSOLUTELY NO SMOKING WITHIN OR NEAR THIS BUILDING

and snow didn't pile a hundred feet of ice atop the hut in the 49 years since Scott built it?[23]

Ever hear of "The Pole of Inaccessibility"[24] (outside *Flying Saucers*, that is?). Well, it's on the maps (some maps) of the North Polar Area. It is supposed to be a land mass between Alaska and the North Pole. In 1925 Jacob Gayer was the lensman on the *National Geographic*–U.S. Navy expedition to find it.[25] All the expedition found was mirages — no land. Jacob Gayer took the first color pictures north of the Arctic Circle, and we wonder if he took pictures of the mirages. Mirages, by the way, are **real** — they are reflections of land beyond the horizon. We wonder what land it could be; the nearest land in that direction is over a thousand miles away. Quite a jump for a mirage!

It is often that this editor makes obvious statements, but maybe this is not one of the times! The implications might not be so obvious. In this issue you have read something of NICAP's challenge about governmental secrecy regarding UFO. In past issues you have read a varied diet of semi-commendation and of criticism for NICAP (which, if you don't know the meaning of the letters is Major Donald Keyhoe's[26] organization National Investigations Committee on Aerial Phenomena).[27] In fact, Mr. Keyhoe has been angry with us several times. Your editor is a member of his organization, paying dues, and getting the bulletin published by the organization. In the main, it does a lot of good work, and the information contained in the bulletin is valuable and

[23] The Hut is 50 feet (15 m) long and 25 feet (7.6 m) wide, and was prefabricated in England, then shipped to Antarctica.

[24] The location, existing for both continents and oceans, that is the furthest from any coastline, and represents the point most difficult to reach due to its remoteness.

[25] Jacob Gayer (1884–1969) was a staff photographer for the National Geographic Society.

[26] See p. 262, footnote #2.

[27] NICAP was a non-profit UFO research organization active in the United States from 1956 to 1980.

helpful. Our major criticism to date of Major Keyhoe has been his continual beating of the drums for interplanetary origin of the flying saucers. Among other criticisms was one in which Kenneth Arnold was involved, and on which we have the most inside of inside information, the ill-advised Armstrong Circle Theatre TV "bust" on UFO, wherein Major Keyhoe was supposedly cut off the air when he began to demand a congressional investigation.[28] We hinted that we felt the whole thing was a "fix." Kenneth Arnold, sensing that the show was to be rigged, refused to go on it, even after being brought to New York for that express purpose. The program, in its presentation, did inestimable damage to the cause of flying saucers, we believe. But in one respect it helped, because to those "in the know," the rigging was too obvious, and only served to make them more fervent in their efforts to find out more about UFO.

This time we want to criticize nobody, least of all Major Keyhoe. We want only to make a point, and we hope to hurt nobody's feelings. The point is very simple: Vice-Admiral R. H. Hillenkoetter[29] was Director of the CIA (Central Intelligence Agency, now headed by Allen Dulles, brother of John Foster Dulles)[30] on January 17, 1953,[31] when a classified CIA-UFO report was drawn up, the complete text of which

[28] In February 1958, Keyhoe was a guest on the *Armstrong Circle Theater* television show, for a segment entitled "UFOs: Enigma of the Skies," in which he departed from the script he had hesitantly agreed to beforehand, and said on national television that the Air Force had three secret documents of which the public was unaware, but before he could elaborate, the producers mostly muted the audio, later explaining that they had censored Keyhoe because they feared a libel suit against the network.

[29] Roscoe Henry Hillenkoetter (1897–1982) was the third director of the post–World War II United States Central Intelligence Group (CIG), the third Director of Central Intelligence (DCI), and the first director of the Central Intelligence Agency created by the National Security Act of 1947.

[30] Don't know the Dulles brothers? Read *The Devil's Chessboard: Allen Dulles, the CIA, and the Rise of America's Secret Government* by David Talbot, and *The Brothers: John Foster Dulles, Allen Dulles, and Their Secret World War* by Stephen Kinzer.

[31] Our research shows that by this date, Hillenkoetter had left the CIA to return to active military duty, and Allen Dulles was then-director.

ch. ends p. 317

Congress has been unable to obtain! As you can see, this is a loaded, double entendre point, because it also asks if there was such a report? Your editor just doesn't know.

Because of Hillenkoetter's CIA background, this editor would like to advance as possibility that NICAP is actually a UFO mouthpiece for the CIA (or whatever agency of government, if such exists, that is responsible for saucer secrecy).

There are several factors in your editor's mind which arouse his thoughtful suspicion, which existed from the first day we received the original prospectus of NICAP. The dues, for one thing, were $100.00. Or perhaps it was membership, we dis-remember — but a lot of money was involved. This was hastily cut down, several times, when the very amount produced no results, and in fact, brought suspicion. Next, in spite of the fact that this editor is not only the **first** flying saucer investigator, but the possessor of the largest private file of information in the world, and the publisher of the only newsstand magazine on flying saucers, and has repeatedly offered to help NICAP, this help has been refused. We even offered to publish their magazine as a section of *Flying Saucers* with absolutely no editing, the entire section to be exactly as prepared by Major Keyhoe, and to pay the entire cost of production with our part being our regular subscription price. At that time this would have left NICAP the larger share of the membership fee, and would have provided the capital he said he needed. But did he need capital then? At first, it seemed, he had money to spend. It was probably his, but it was soon gone. We privately wondered if it was his?

Next, appearing on two national television hookups (Mike Wallace's odious "pin the tail on the donkey" show was the other),[32] Major Keyhoe succeeded only in making the case for UFO look very weak.

---

[32] On March 8, 1958, Keyhoe appeared on *The Mike Wallace Interview* on ABC and spoke about the details of the *Armstrong Circle Theatre* censorship, which he blamed on the Air Force rather than CBS.

To be entirely fair, nobody else has made the UFO subject appear very strong, and that includes us. Not in an official governmental way, that is. We've convinced a lot of private citizens, but have gotten nowhere in overcoming the secrecy and ridicule blanket that is so obviously being thrown over the whole subject. In that, Mr. Keyhoe can hardly be condemned only regarded in mutual sympathy.

But one accusation we made, which was the only one that aroused Major Keyhoe to a public denunciation of us, was the fact that NICAP's board of governors and advisors was loaded with military personnel. Now, in the interest of adding to the record, we make our point regarding Vice-Admiral Hillenkoetter. He is not only military (retired) but also **secret** intelligence (retired)!

We may be all wrong, and the whole of NICAP's membership is sincere and truthful. But it is a **fact**, and one that cannot be ignored without great danger of doing serious harm to the effort to get the truth of UFO, that it is **impossible** to take things said by Hillenkoetter as gospel, because of the background of the man. He was head of the CIA, and he's like a wife who cannot testify against her husband — his testimony is legally prejudiced and unacceptable. When he talks of secrecy, it must be remembered that he was secrecy in the past, and **top man**.

Let's not be lulled into ceasing our own efforts to get congressional investigation into flying saucers by the supposition that NICAP and Hillenkoetter are on the job and that therefore everything possible is being done, and all is well.

Of course what we have said has little point, other than the fact that is already obvious, NICAP is still loaded with ex-military personnel, and to people with suspicious minds (your editor has the worst possible case!) whatever they say should be examined carefully.

Let it not be misunderstood — we believe NICAP is a good thing, and we think everybody interested in finding out the

ch. ends next p.

truth about UFO should be a member, and get that publication. It contains vital information. It is valuable. And we may be wholly wrong, and the implication of the military (and intelligence) personnel preponderance in NICAP is because each and every one, now retired, is making an effort to overcome an evil he could not combat while actually in the service! Accordingly, we print herewith the address of NICAP, for the benefit of all who wish to join it. Write to: NICAP, 1536 Connecticut Ave., N. W., Washington 6, D.C.

The March, 1960, bulletin carries no membership fee, nor subscription price, so we can't give that to you, but you can write for the cost if interested — and you should be.

It would seem that there are more persons in this strangely shaped world who are in doubt about, the poles. Now Swedish polar experts are challenging two American North Pole claims. They doubt whether Admiral Richard Byrd ever reached the pole on his famous flight of May 9, 1926. And they say it's unlikely that Admiral Robert Peary was the first man to reach the pole. Peary is credited with reaching the pole on April 6, 1909. Professor Gosta Liljequist[33] of Uppsala University[34] says that weather conditions on the day of Byrd's flight made it unlikely that he could have reached the pole and returned to base in the 15 hours and 30 minutes claimed. And Dr. Valter Schytt[35] of Stockholm University says of Peary: "It is unbelievable that such an old man who could not ski and who traveled on foot could cover between 100 and 120 kilometers a day over rough polar ice."

[33] Gösta Hjalmar Liljequist (1914–1995) was a Swedish meteorologist who took part in the 1949-1952 Norwegian-British-Swedish Antarctic Expedition (NBSAE).

[34] Founded in 1477, Uppsala University is the oldest university in Sweden, and the Nordic countries, still in operation.

[35] Stig Valter Schytt (1919–1985) was a Swedish glaciologist who also took part in the 1949-1952 Norwegian-British-Swedish Antarctic Expedition (NBSAE). Schytt Glacier in Antarctica is named in his honor.

No less an authority than Bernt Balchen[36] agrees that Peary could not have reached the pole.

In our next issue, we will try to have a very exciting article concerning the mysterious "Pole of Inaccessibility," and Admiral Byrd's mysterious "land in the sky" statement. So seriously was this taken that an actual expedition was sent to photograph it.

We will also further our "case for a hollow planet" with a very intriguing article which will sum up some of the evidence that is readily available which points to the possibility. We have in preparation another article which culls from the Bible a surprising amount of testimony that the writers of the Bible believed the Earth to be possessed of mystery lands and openings at the poles. To the believer in the Bible, this article will be quite a staggering one. Your editor believes in the Bible, and in fact, has studied it since boyhood, intrigued by what it says that nobody else seems to realize it says. The science hidden in the Bible is striking in the light of this century's discoveries. The Bible says the Earth is hollow, and tells why it is (or was) not possible to get into it.

A very important article to come will contain an interview with (name withheld for reasons that will become obvious when the article is published!) which will give the facts about Byrd's 1926 polar flight!

If you missed the December issue of *Flying Saucers*, in which all this inner earth material began to appear, and the February issue in which it was continued, we still have copies available, at 35¢. You may regret not having a copy as time goes by and future issues of *Flying Saucers* get you intrigued! —Rap.

---

[36] Bernt Balchen (1899–1973) was a Norwegian pioneer polar aviator, navigator, aircraft mechanical engineer, and military leader.

FLYING SAUCERS

The Magazine of Space Conquest

AUGUST, 1960

WISCO

35¢

The HANOVER "BALLOON"

Did Flying Saucers "Watch" The Wallops Island Firings?

The POLAR MYSTERY AREA
—WHAT THE BIBLE SAYS ABOUT IT!

AIR FORCE RIGHT ON KILLIAN "SAUCER"?

Editorial

Flying Saucers #16, August 1960

A needle is quite a simple thing; for sewing, that is. When it comes to the kind of needle we put into a compass, that's an entirely different thing! Maybe you've found it out, too, if you've been doing any really serious investigation into our theory on the strange condition of the Earth at the Poles. But if you haven't, let's take a peek at the mental gyrations this poor editor has been going through on this one simple mechanical device. The question of the compass became very important when some of our readers made some flat statements about navigation in the Polar area. We even got letters from Polar Navigators (aircraft, that is) on how we were all wet.[1] We looked into the matter, and decided that we would make no Polar flights with these fellows as navigators! Not that they don't do a good job, but they do it **all on one side of the pole**. If we were going to request a flight across the pole and **into the other hemisphere**, we'd rather do our own navigating, because we just couldn't trust these fellows **once they violated the cardinal rule** of "playing it safe."

Before we get into the interesting material behind these

[1] To be "all wet" is to be all wrong.

innuendos of ours, let's cut the compass thing short with a few flat statements of our own.

1) The compass is worthless for navigation within a thousand miles of the Pole, and in fact, it's worthless almost anywhere. It just gives you a **general idea** of which is North or South, and if you really want to know where you are and where you are going, you have to "shoot the sun" with a sextant.[2] Further, have any of you asked yourself which direction the compass needle points at the Equator? Well, where does it point? The answer is, we're hanged if we know!

2) The gyrocompass is the instrument used instead, for navigation. It is much more dependable than the magnetic compass, and it is best to use both together, so as to get the most out of either.

But let's make what seemed to us at the time a fantastic discovery when we read it in the July 1959 *National Geographic*: "*Nautilus* had made an exploratory cruise under the ice pack in the fall of 1957 and discovered there were problems. Her gyrocompasses had failed . . ."[3] And again: "By the next afternoon we (*Skate*) were within 150 miles of the Pole, and our regular gyrocompasses could no longer tell us which direction was north."[4] Well, there you have it. As you approach the Pole, your vaunted[5] gyroscope **can no longer tell which direction is north**! So, as an aside to those pilots who wrote us such sarcastic and uppity letters, you can just go jump in the Arctic Ocean with your gyrocompass, if you think this editor is going to trust you to use it in taking him **anywhere** near the Pole!

The foregoing are facts. No argument about them. So what

---

[2] An astronomical instrument used to determine latitude and longitude at sea by measuring angular distances, especially the altitudes of sun, moon, and stars.

[3] Calvert, J. F. (1959, July). Up Through the Ice of the North Pole. *The National Geographic Magazine* 116(1), p. 6.

[4] Ibid. p. 16

[5] Excessively praised or boasted.

did *Nautilus* and *Skate* use in navigating under the polar ice? Well, they invented a new use for a device originally designed as a guided missile gimcrack[6] — the "inertial guidance system"; they adapted it to submarine navigation. In short, this system senses the Earth's rotation and charts its speed. Delicate instruments feel the direction of the motion resulting from the Earth's eastward spin — thus they tell which direction is East. They sense speed as well as motion, and since the rotation of a point on the Earth's surface decreases as one goes toward the Poles, the inertial gadgets can sense their distance from the Pole.[7] Theoretically (our word, not theirs, because theirs is spelled positively) when the gadget registers no motion, the location must be the only point on the Earth's surface where there is no eastward motion, the Pole — the exact Pole. Our reason for saying theoretically is one that will get us plenty of arguments; so we'll merely say that theoretically, we don't feel that inertia, gravity, mass, electromagnetism are at all understood, and actually, we are not at all positive as to their relationship with each other, and whether or not one is effected by the other. In short, is inertia **really** a constant? In a previous issue of *Flying Saucers* we told you how the Russians had gotten the peculiar idea that retro-rockets might not be necessary on landing on some planetary bodies, because the gravity-field was different than the one peculiar to the Earth. Also, they had the idea (now widely held, as we shall show in future discussions) that inertia **beyond gravity** may not exist, and that therefore, once beyond the clutches of gravity, a spaceship can be propelled by an **ion beam**, which has hardly more energy

---

[6] Something that is useless with no value.

[7] The system uses sensors to track movement and orientation without external references, relying on accelerometers to measure acceleration and gyroscopes to measure rotation, using these measurements to calculate position, velocity, and orientation. The self-contained system is valuable in situations where external navigation aids like GPS are unavailable or unreliable.

ch. ends p. 328

than an ordinary flashlight. The Russians seem to suspect that gravity does not decrease inversely with the square of the distance, but cuts off abruptly at a sort of "boundary." Wonder where they'd get such a queer idea? And still hit the moon precisely, while we miss it by 79,000 miles!

Yes, you've guessed it by now — we think *Nautilus* and *Skate* reached a point where the inertial guidance system showed no eastward motion of the Earth, but **it is merely an assumption** that this lack of motion means the point reached is the **geographic** North Pole. All it means is that they reached a "failure point" where the inertial guidance system no longer is able to determine motion, just as the gyrocompass, at a distance of 150 miles from the Pole, is unable to decide when it is being swerved and not being swerved from its axis of rotation! If anybody had said that a gyrocompass could be turned on its axis and **not register** that turn, he would have been laughed out of the halls of science. Yet, when *Nautilus* went to the Pole, they found out that this was precisely what happened, and the gyrocompass was useless for navigation. No loud screams from the halls of science — just a flat admission in the *National Geographic* that it is so. We challenge science to explain **why** to us! When they do, we will be able to tell them why the inertial guidance system is subject to the same failure.

By now, you must be as confused as we are about compasses. Their pesky needles are sticking us in a tender place, suggesting that the only smartness we can claim is the "smart" of the prodding needle. Perhaps a glance at the map showing *Skate*'s Polar gyrations will give us some clue to a way out of our confusion. If the inertial guidance system is unreliable in any way at all, it should show up in the *Skate*'s maneuvers. And so it does! *Skate*'s explorations very carefully cover only 15 of 18 parallels of longitude **in one hemisphere**. Perhaps they could have navigated in the other 21, but they would have **gone around** the Pole to do it. **Not across!** They **went**

around the Pole on every maneuver, except the one where they went **to** it. And this peculiar fact leads one to wonder if the path of the *Skate* was not necessarily so circumspect because of **inability** of the inertial guidance system to guide them in anything but an area **this side** of the point where it stopped registering motion!

We are in a decidedly weak area of discussion here, because it doesn't necessarily mean that *Skate* couldn't go where we think it should have gone. It could mean it was under orders to go only where it actually went. But it is true that it never made a move toward that mysterious direction we are forced to call (as Admiral Byrd did) **beyond** the Pole. Why did they not proceed in a direction which would cause the inertial guidance system to report motion in a westward direction? That would have been **beyond** the Pole! But they carefully stayed in an area where the inertial guidance system indicated eastward motion. Or will they tell us that there is no such thing as westward rotation of the Earth? To a layman like ourselves, we can picture the *Skate* proceeding toward the North Pole, watching its gadget register a constantly decreasing eastward motion of the Earth, until at last we jubilantly cry "The Pole has been reached"; and we can picture the engineer, unaware of the cry in his engine room, blithely driving the ship forward, and the gadget beginning to register motion in the **opposite** direction! How could it do otherwise? The Earth is rotating the other way, in relation to the inertial guidance system's original heading, on the other side of the Pole. So, it should register westward motion! How about navigating from there? Chart a course for central Siberia! Don't back up and square the gadget with the east, as it should be! Plunge on. See if you can navigate in reverse! Afraid to get lost?

Now we come to that mysterious man, F. Amadeo Giannini, P.O. Box 695, New Providence, New Jersey. We give you his address because we think he should get some

ch. ends p. 328

attention! He should have an opportunity to explain himself to any one of our readers who cares to check him personally, and not just take our word for it.

Such a person is reader Richard Ogden, 1233 Ninth Ave., West, Seattle 99, Washington. He wrote Mr. Giannini on April 25, 1960. Here is what Mr. Giannini replied, word for word:

Replying to yours of the 13th instant: Please be advised of the following: The standard reference works, dealing with the late Rear Admiral Richard Evelyn Byrd's Antarctic exploration do not mention the 1,700 mile flight over land beyond the North Pole point of theory, in February 1947.

And there is no question about Rear Admiral Byrd's Antarctic venture of 1946. But that recorded Antarctic exploration of 1946 can in no way detract from the U. S. Naval task force flight over land beyond the North Pole in February 1947.

This author cannot be held responsible for political whims and any and all political and militaristic mis-representations of fact imposed upon an otherwise intelligent, but conveniently considered most unin-telligent and gullible public of the year 1947 or 1957.

Hence your interest may be satisfied by *The New York Times* accounts of December 1946 through February 1947. And, if they will, responsible agents of the U. S. Naval Intelligence Office, Wash., D.C., can provide additional satisfaction.

Please be assured that your interest is deeply appre-ciated, as is the interest of Mr. Ray Palmer who has purportedly "reviewed" this author's book. And, in that another source has informed this author of Mr. Palmer's enthusiastic designation of liar, a copy of Mr. Palmer's "review" would be welcome. For, in the modern order of surpassing lies and liars, even the liar designation would hold certain merit for this author.

That ends Mr. Giannini's statement. Note that it repeats what he said in his book — that there was a flight over the North Pole in 1947. It infers that there is "political and militaristic misrepresentation of facts" concerning this flight. Then it goes on to say Ray Palmer reviewed his book and called him a liar. Neither is true. We mentioned his book, even quoted from it, but we didn't review it. If we had, we would have wound up talking to ourselves! Note, however, that Mr. Giannini stakes his verity on *The New York Times* accounts of December 1946 through February 1947. In short, if such a North Pole flight was made, mention of it is in the *Times* during that period. Is it? The answer is no. All about South Pole flights, but not North Pole flights. For which research we thank Mr. Ogden, and we courteously point out that if anybody is calling Giannini a liar, it is Mr. Giannini.

But then Mr. Odgen goes on enthusiastically to state that all this not only disposes of Mr. Giannini, it also disposes of Ray Palmer and his theory of something strange at the Poles. No equivocation — just plain and simple "out with it," says Mr. Ogden. To your editor this is tremendously interesting. As we have pointed out before, our original covering of this mystery in the December 1959 *Flying Saucers* gave a wealth of information including the now false Byrd North Pole flight. But it did **not** give a single word about the 1947 South Pole Flight. At the most, all that could be claimed is that we got the Poles mixed up. We also pointed out later that we deliberately used the Giannini version because it was the **only** way to smoke out a "classified" North Pole flight, if it existed. What it smoked out is Mr. Giannini, and in the process, poses a truly fantastic mystery. **Why** did Mr. Giannini make his supposedly stupid mistake, which he **still** adheres to, but leaves himself clear to be lightly dismissed by setting up

ch. ends p. 328

an "out" in the columns of *The New York Times*. He wants to be lightly dismissed. We don't intend to let him be lightly dismissed!

We've read Giannini's book again, and we can only say that throughout the book, he belabors the point. The entire book is hung on that one fictional flight! If Mr. Giannini was familiar with the *Times*, as he evidently is, he **knew** his whole book was based on a weakness that could not fail to destroy his theory, and entirely obviate[8] any reason for writing it at all! He deliberately wrecked his own book. Why?

Oh, we know why; and the why is glaringly evident in Mr. Ogden's delightfully final dismissal of the "whole thing," in spite of the fact that the smallest supporting item for our theory was the Giannini episode. A whole convincing mass of material that **cannot** be dismissed appeared in that same December issue of *Flying Saucers*, and in the February 1960 issue, and in the June 1960 issue, and in **this** issue, and in issues to come, which Mr. Ogden chooses to bypass with stealthy steps. Again, why?

No, Mr. Ogden, our theory isn't dead. It has been remarkably enhanced. Mr. Giannini, if we may be permitted, may not be a liar at all! He may have done what we must call "Palmer's opposition in this Polar Theory business" a great service. For the sake of our ego, let's go ahead and call everybody who opposes our theory as "Palmer's Opposition." Mr. Giannini did something in writing that book, and we'll be hanged if we can put our finger precisely on what it was he did!

Recently, geography has become a number one subject for study by this editor. He has gone "back to school" to find out what it was that he missed, and is quite staggered to find out he missed so very much! All our old concepts have been smashed. All the little pet beliefs we had about the Earth and

---

[8] Remove; render unnecessary.

where everything is are shattered. We used to think, as we sat on the shore of Lake Michigan and watched the ore boats and car ferries go "hull down" over the horizon, that we were "checking on our geography teacher." It was wonderfully stimulating to observe the phenomenon that proves the Earth is round! But now we'd like to ask our readers to do a little checking on this very thing, with one exception. After you've watched the ship go down beyond the horizon until only the smoke (not even the smokestack) is visible, with the whole ship down out of sight behind the curve of this round Earth, set up a 50 power telescope and take another look.[9]

That's all we ask.

We want to know what you see. You can bet that as soon as we get a chance to go to the lake shore, we're going to cart our telescope with us and watch a ship go hull down with our naked eyes until it has disappeared, and then we're going to use the telescope.

All we should see is the smoke coming up over the horizon.

Most of you will agree, won't you?

But will some of you do it, just the same?

And tell us what happens?

Nothing will happen. It couldn't! All the telescope will do is bring the smoke, and the horizon, nearer, make it more plainly visible than to the naked eye. We're absolutely sure of that, because the geography books tell us the Earth is round, and this hull-down thing proves it.

You see, this editor has to be reassured now, with all this amazing mass of evidence coming in that something is awry with the Earth and its shape. He's getting a little bewildered, and he could use a little "down to earth" reaffirmation that everything hasn't gone topsy-turvy.

How many of us have gone through life, reading the textbooks, faithfully following their dictums, fatuously sitting on

[9] Telescope power, or magnification, is how much larger an object appears through the telescope compared to the naked eye.

ch. ends next p.

the seashore, declaiming that "seeing is believing," and never giving the matter another thought? How many of us bother to think everything through, rather than accepting somebody else's thinking? How many of us actually have no proficiency at thinking, because we never think? Is it too hard? Or is it a waste of time to go through something another mind has already worked out? What about doing such thinking only as an exercise? If exercise strengthens muscles, won't thinking strengthen the brain? Is it so bad to have a strong brain? Thinking, like exercise, should begin with the easy things. You don't go in cold and pitch nine innings—you warm up in the bullpen first. You think about simple things. Like the hull-down proof that the Earth is round first with the naked eye, and then, quite silly of course, with a telescope. But is it silly? Isn't the idea of enhancing your vision in this test of the roundness of the Earth a good one? If the naked eye is an instrument that proves an important point, isn't it a good idea to make doubly sure, and add a telescope to the naked eye?

Maybe all we're doing is trying to trick you into a little thinking. But isn't your curiosity roused? What **do** you see when you add a telescope? Has anyone ever done it? And wouldn't it be nice to be able to report that even with a telescope things are still on the up and up in the geography books, and we didn't waste 5th grade after all!

Mr. Ogden dismisses things too easily — and maybe a little simple exercise such as we've just described would teach him not to ignore hundreds of little facts because they are overshadowed with an **obvious** one. Mr. Giannini's book may be obviously in error, but Mr. Giannini himself is not at all obvious! —Rap.

Antarctic Expedition
Called Big Success

Cruzen, the task force chief, said naval personnel on the expedition got a "real insight into the type of problems you have to risk in polar regions, and that was the primary objective—to find out what our problems are."

Rear Adm. Richard E. Byrd's flight over the south pole obtained good photographic coverage of new areas, especially in penetration beyond the pole, where a plateau was found to extend flat and featureless, Cruzen continued.

0 1000
STATUTE MILES

90°
**SOUTH
AMERICA**

Pacific Ocean

Cape Horn

Franklin D. Roosevelt Sea

Palmer Peninsula

Weddell Sea

Wrigley Gulf

Ellsworth Land

Filchner Shelf Ice

Marie Byrd Land

60

Little America

Ross Sea

180°

SOUTH POLE

0°

Queen Maud Land

S. Victoria L.

ANTARCTICA

Wilkes Land

60°

Adelie Coast

Knox Coast

Antarctic Circle

Indian 90° Ocean

Associated Press Wirephoto.

THESE AREAS on South Pole maps will have to be revised because of discoveries by Admiral Richard E. Byrd's expedition: 1, the Walter Kohler mountain range (peak symbols) runs north and south whereas present maps show it east and west; 2, a good-sized bay was found running inland on eastern side of the Roosevelt Sea; 3, a peninsula was found west of the Adelie Coast where maps heretofore showed a bay, and 4, a new bay was charted on the Knox Coast area of Wilkesland.

FLYING SAUCERS

The Magazine of Space Conquest

WISCO

35¢

NOVEMBER, 1960

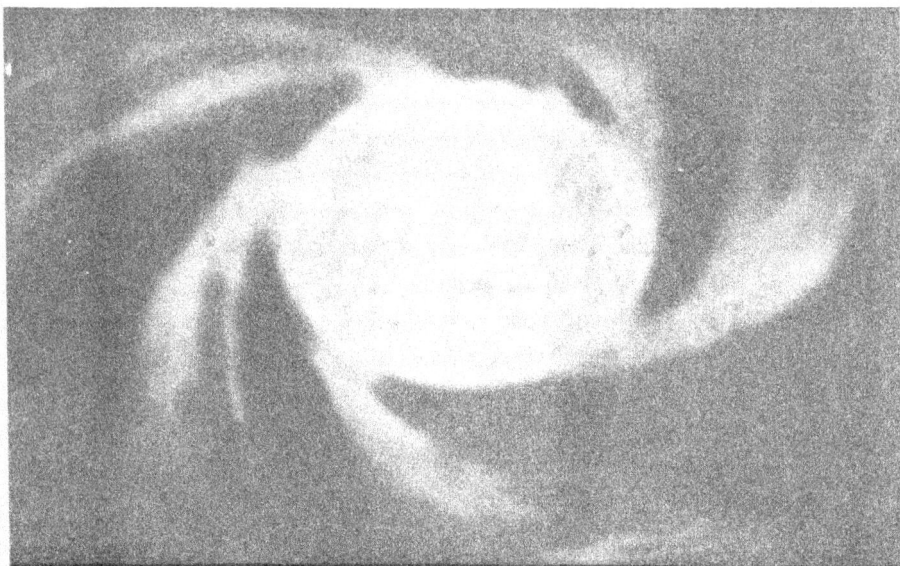

SPANISH U.F.O. STILL UNIDENTIFIED

WHY DO POLAR ROCKETS GET LOST?

A mystery that is costing the U. S. taxpayer uncounted millions!

EARTH'S "CENTER OF GRAVITY" — UP OR DOWN?

Did U. S. — French governments bottle up baffling discovery?

1896 SKY SHOW - Sensation of a century!

Editorial

Flying Saucers #17, November 1960

An editorial is usually the place where the editor gives his opinions, presents the readers of his magazine with a personal picture of what goes on in his own mind, and draws his own conclusions of the articles and information printed in the magazine proper. This issue, we intend to open with a mere presentation of facts, and refrain from drawing any conclusion, or offering any opinion. We will leave it strictly up to the reader to decide for himself. When you've done so, don't later infer that this editor said anything one way or the other about the list of facts he presented.

The facts are these:

1. With our December issue we began publication of a new trend of thought, basically that all is not as we have been brought to believe about this Earth insofar as its physical nature is concerned. We presented it "cold" because we had a hunch there just **might** be opposition to it that would effect our sales. In short, we suspected there might be censorship.

2. On January 19, 1960, we received word that our printer might not be able to continue printing our magazines, due to a very bad situation that had arisen.

The trouble began with a paper shortage brought about by a sudden purchase of the paper supplier by another firm which arbitrarily decided to cease making the particular type of paper we were using. As a result, the February issue of *Flying Saucers* was actually delivered to the newsstands **two months late**, thus almost destroying our period of on-sale because it is a fact that newsstands remove from sale in February all magazines dated February, and many magazines dated March (as out-dated). A further result was the complete skipping of the April issue. The next issue was dated June, and in spite of this lag, the June issue hit the newsstands **in June**, and suffered the same fate — no sale period. The next issue was August and managed to get on sale by July 15, and suffered an almost equally serious setback in on-sale time.

3. More than 3000 copies of the February *Flying Saucers*, consigned to us via truck, vanished into nowhere. No trace of them has ever been found, not even a waybill.

4. 50 copies shipped to a California distributor were not delivered (because the distributor himself vanished into thin air — also without paying us over $600 he owed us): yet these copies, on which labels clearly define that if undeliverable return postage is guaranteed, were not returned to us. In spite of this, the magazines could not be found by the U.S. Mails anywhere in their possession.

5. The **entire** state of Pennsylvania's single subscribers received only **empty** envelopes. No magazine was enclosed. We had to ship out a second shipment to all our Pennsylvania subscribers.

6. Of more than 8000 copies shipped to California newsdealers, only 3400 were ever delivered. 4600 copies vanished into a limbo that is inscrutable. None were

ever returned to us, none can now be found in any post office, none were delivered.

7. While it is a simple matter to put the plates back on the presses, **this one time** it was impossible, because **somehow** the plates had been so badly damaged by unknown agencies that they were useless for a reprint. The printer was unable to explain how the damage occurred.

8. A number of distributors abruptly canceled their orders and refused further to handle the magazine, in spite of the fact that sales percentages were remarkably high.

9. The habit of the newsdealers in paying bills promptly and in full almost to 100% within 30 days of billing suddenly reversed itself, and only 15% of the accounts were settled up, leaving a large unpaid balance.

10. The Discount Corporation handling the printer's accounts, suddenly, and after five years of prompt monthly payment of our printing bills, threatened to take court action to collect a balance of $6700.00.[1] (We mailed them a check the next day in full, and have since refused to allow our printer to discount our bills, and have paid the printer **cash on delivery** ever since. To do this, we mortgaged our home. But one thing is sure — nobody ever threatens Ray Palmer without risking the hottest temper in the business. Nor do they ever do business with us again!)

11. The postmaster general's new "regulation" which designates unsold newsstand copies as "Free" copies, and therefore not eligible for second class entry, in actual effect will destroy every "little" magazine in the country. However, this was designed to bring the percentage of sale below a 70% figure arbitrarily

[1] Adjusted for inflation, that's approximately $73,000 in 2025 dollars.

ch. ends p. 339

set up by the big magazines, and in our case it fails utterly, since our sale, under our own distribution efforts, is well above 90%.[2] True, *Flying Saucers* does not appear on thousands of newsstands, but on those where it does appear, it reckons its newsdealers as the most honest and trustworthy and hard working in the country.

12. This issue is dated November 1960, instead of October, as it should be. This is designed to restore us to full on-sale time, and to insure prompt issuance of *Flying Saucers* every two months from now on without delay.

13. Except for a very tight situation regarding available operating capital, our troubles seem to be over. We've licked everything Fate could throw at us (not the magazine *Fate*,[3] but that fickle old lady who thinks she runs things), and we're stronger than ever.

Now, make what you want of that list — it's all in the past, and meaningless insofar as we are concerned. Our survival has proved the soundness of the lessons we learned in the depression of 1954.[4] We're rather proud of the fact that at this date we have two magazines, and are engaged in printing the first issues of two more! One of them is *Hidden World*, which will present, for the first time anywhere, the **entire** Shaver Mystery, with all its evidence, in a totally factual manner, stripped of all fictional trimmings. The

[2] Second-class mail (now called Periodicals class) was reserved for publications that met specific criteria, and offered reduced postage rates for newspapers and magazines. In November 1960, the Postmaster General issued a new interpretation that unsold newsstand copies—those returned or not purchased—would be considered "free" copies. Reclassifying them as "free" reduced the percentage of a publisher's paid circulation (one of the required criteria), which meant the publisher could lose eligibility for second-class mailing privileges, effectively increasing their mailing costs and reducing both their profit and distribution.

[3] Still in publication today as the longest-running magazine devoted to the paranormal, *Fate* was co-founded in 1948 by Palmer and Curtis B. Fuller.

[4] The post-Korean War recession that occurred in the United States from July 1953 to May 1954.

number of pages in this new magazine (it isn't a magazine at all, but a book!) will be 192 (three times that of *Flying Saucers!*), and will sell on a **subscription basis only**. The book will be printed on permanent, good-quality paper, with heavy cover.[5]

Our decision to publish this type of "book" is in the way of a prophecy. We predict that sales of small magazines via the newsstands will become a thing of the past, for a number of reasons:

1. The inability of the newsdealer to keep his handling costs within a figure that will allow for small accounts; he will be able to handle only the big publications at a profit.
2. Pressure by the big magazines in the way of competition for newsstand display space.

There is no question as to who will get the space. And a magazine that cannot be seen, cannot be sold. We further predict that small publishers will be unable to raise prices on newsstand periodicals to cover their own costs, because they do not have the advantage of the big publishers whose advertising revenue is their means of support while their product actually sells at the newsstand for **less** than their production cost. The big publisher loses money on every magazine he sells, but pays the deficit and makes his profit on advertising alone. Because newsdealers cannot possibly sell a 64-page magazine (such as ours) for what is actually necessary to make it profitable (50¢ or over) when he sells the big ones for as little as 20¢, small publishers dare not raise their prices higher than the 35¢ they have now reached (although some few have charged 50¢, but are rapidly going out of business as sales drop).

Thus, we feel that it is the 100% subscription magazine that will be the only hope of the small publisher. In fact, for

[5] In all, 16 issues of *Hidden World* were published quarterly from Spring 1961 to Winter 1964.

ch. ends p. 339

years we have urged our readers to subscribe. As a result, there are many thousands of people who were avid readers of anything we published, who today believe we went out of business years ago! How could they know that when the magazines stopped appearing on their newsstands, it was not because they weren't being published? Thus, they lost contact, and were deprived of something they really wanted. We don't know how many times we've said that if only 50% of our readers were to subscribe, we would be able to put out a magazine **twice as thick** and **every month**. It has not been because we don't want to publish that much and that often, it is because you, the readers, don't want us to! You say you do, but somehow you are so short of money that all you can raise at any one time is 17½¢ per month! The task of getting together $4.00 for twelve issues is beyond your financial capabilities! It must be that this is so, because almost invariably readers who write us about the frequency of our appearance bemoan the fact that it is not oftener.

Thus it is that with *Hidden World* there will be no newsstand sale. You'll have to subscribe to get it, and when you do, you will have concrete evidence before you of what *Flying Saucers* could be also! You might ask, why then do we not make *Flying Saucers* a subscription magazine, on the same basis? The answer to that is simple — we can't let our thousands of newsstand readers down — we've got to make their favorite reading matter still available to them. They won't or can't subscribe, but we won't disregard them. And finally, we believe firmly in everything we publish, and in our purpose in publishing it, and we want to reach everybody we can — because it may be of value to them. It is only that we see the handwriting on the wall, and when the final blow falls, we will have established ourselves on a safe subscription basis with two new magazines, and not all will be lost.

Times are changing. The methods of doing things are changing. Our American way of life is being attacked in many subtle ways — and the attack on our liberties, our privileges, our freedom of speech, our freedom of expression and exchange of ideas, by the **dollar** is the most subtle and dangerous of all. It is becoming, literally, too costly to think for ourselves. We have wonderful center-drawer refrigerators and color TV, but in the range of the 35¢ magazine we are rapidly being stripped of the wherewithal to both buy and produce such magazines. Thus, it seems logical that the thinking man's magazine be modernized, dressed in a bright new garb, put on a dollar pattern compatible with the times, and our freedom of expression restored to its rightful availability.[6]

We'll make you a promise — if each and every one of you who is a regular newsstand reader will subscribe right now, and promise to remain a subscriber, we'll put *Flying Saucers* on a monthly basis, and restore it to 96 pages. And we'll still be able to put it on newsstands, and reach a brand new audience just as large as the one lost to the newsdealer by those who subscribed — because the magazine you previously bought will still be there for the customer who has not seen it up to now![7]

There are no doubt those who will accuse Gray Barker of sensationalism in his article in this issue.[8] We admit that it is one of those things that cannot be substantiated. However, there seems to us to be room in *Flying Saucers* once in a while for the unsubstantiated, even the article that can be called "imaginative" if not worse. But let it be known that neither Gray Barker nor your editor present this material as anything more than provocative. Let it do whatever it will to

[6] We Heathens believe the rhetoric that Palmer was peddling in this paragraph in 1960 just goes to show that the times don't really change . . .

[7] Making good on his promise, starting with *Flying Saucers* #19 for May 1961, Palmer reduced the magazine's dimensions to digest size, increased the page count from 64 to 96, and began printing the magazine himself. In the Editorial of that issue, he thanks the readers for answering his call to renew their subscriptions en masse.

[8] Barker's recurring column was "Chasing the Flying Saucers."

ch. ends next p.

your thinking. Sometimes even a false note jars a complete picture into place — we see what we failed to see previously, because of an incongruity. We wonder what incongruity you will find in this particular article?

Naturally, in "Why Do Polar Rockets Get Lost?" you will find a few more factors linking our Polar Theory to the saucers. But the prime purpose of this article is a tremendous curiosity we've developed since we found out gyroscopes don't work to guide the submarine to the pole. Why not? Maybe the rocket men, facing the problem of the lost rockets, will give us some answers we find devilishly hard to pry out of the physicists. We've written a whole string of letters to prominent physicists at large universities, and we must have driven them to drink, because thus far, not one has answered our challenge to read the July 1959 issue of *National Geographic* and tell us why the statement that the gyrocompass is useless within 150 miles of the pole has not been challenged by these men of science. Perhaps inability to answer is so embarrassing that it engenders silence.

In "Earth's Center of Gravity — Up or Down?"[9] we are calling for some duplication of the tests mentioned, and we think it's important enough now, in the light of man's attempt to reach into space, that the facts about gravity, space, geometry, and physics be attested by actual field experiments such as those undertaken at the turn of the century and then discarded because "the value of the franc will remain unchanged." The franc's value has changed, drastically, several times since then and it's about time we found out if plumb lines do what they are said to do. Just because men of mystic-mind have seized upon such things as this to bolster their own theories (namely Koresh and his belief that he

[9] While it doesn't pertain, exactly, to the topic at hand, so crazy is this article that we Heathens have reproduced it on our website, which can be found linked through our *Worlds Beyond the Poles* product page. We believe it's well worth the read!

was Christ come the second time),[10] let us not abandon vital experiments that today loom as vastly significant. It is vital that we either prove or disprove this 60-year-old experiment, and find out if it is true that the center of gravity is an illusion, or at least not where we think it is. If man is to escape gravity, he must also escape any fallacies concerning it. We don't say that it is impossible that those early experiments were in error — we say that it can be true that they are **not**!

Thus far, to our amazement, we have unearthed **six** books written by serious men who believed the Earth is hollow, with openings at both poles. We don't know whether to be flattered or dismayed that our thinking has been far from original. Obviously the evidence at hand has disturbed others, too, and there has been much thinking through the years about this mystery.

One of the books has had single sentences, whole paragraphs, and even whole pages removed with a razor blade. The purpose of removing a single sentence baffles us, particularly since the sentence removes something vital from the context.

For some reason, saucer researchers are hinting that there is a current "lull" in saucer sightings, and that nothing at all is going on. We can certainly deny this. Our reported sightings by readers is as heavy as ever, and our files continue to grow. During all this "lull" (perhaps in regard to the newspapers only), it may be that we are getting the more legitimate report! —Rap.

---

[10] Cyrus Reed Teed (1839–1908) was an American physician and alchemist. In 1869, claiming divine inspiration, Teed took on the name Koresh and proposed a new set of scientific and religious ideas which he called Koreshanity, including the belief in the existence of a concave Hollow Earth cosmology, positing that the sky, humanity, and the surface of the Earth exist on the inside of a universe-encompassing sphere.

FLYING SAUCERS

The Magazine of Space Conquest

WISCO

FEBRUARY, 1961

35¢

SAUCERS NOW PROVED!

Both Brazil Navy and Brazil Congress Report POSITIVE PROOF!
Photos taken at Island of Trindade aboard IGY ship

"BYRD DID MAKE NORTH POLE FLIGHT IN FEB. '47!"

—Giannini

"THE FLYING SAUCERS ARE HOSTILE"

—George D. Fawcett

"Byrd Did Make North Pole Flight in Feb. 1947!"

Flying Saucers #18, February 1961

Although the following consists of a series of letters between Richard Ogden, a *Flying Saucers* reader who took it upon himself to query Mr. Giannini, author of the mysterious book *Worlds Beyond the Poles*, and Mr. Giannini, we present it as an article, in which the editor intersperses his own comments. Undoubtedly this will lead to still further action on the part of Mr. Giannini, whose place in this weird question of the mystery surrounding both Poles must most certainly be settled in public in the pages of this magazine. Thus, we invite Mr. Giannini to use all the space he needs in our pages to present his case—which as we see it is to produce the proof with which he backed up his book and the information concerning Byrd that he has presented.

The question being placed before the readers of *Flying Saucers* at the present moment is not whether or not the Earth is hollow and that openings exist at both poles, or whether it is otherwise shaped and extends into space at both poles (Giannini's theory), but whether or not Admiral Byrd actually flew over the North Pole in February 1947, and whether or not he "penetrated 1,700 miles **beyond** it" into an unknown land area which does not exist on present-day maps. Therefore, we present the following letters as evidence, and we will let both of them stand on their own merits. Mr. Ogden's letter is actually a composite of two letters, which we have re-arranged as one letter, omitting extraneous comment. If Mr. Ogden feels we have distorted his meaning, he is welcome to correct us. Mr. Giannini's letter is presented verbatim:

1233 Ninth Avenue West

Seattle, Washington

August 2, 1960

Mr. Ray Palmer

Flying Saucers

Route 2, Box 36, Amherst, Wisconsin

Dear Ray:

On July 23rd I wrote another letter to Mr. Giannini asking him to give an account of himself about Admiral Byrd and what *The New York Times* reports. I asked him your question about why he tried to destroy his theory by including a fictional flight by Byrd to the North Pole in 1947. I told him: "I am out to do just one thing: **get the facts**. I will play no favorites. If this thing is a hoax, I will expose it. If there is some truth behind it, I will bring that out too."

Well, it seems Mr. Giannini doesn't give up; and I rather suspect you will be printing an apology to Mr. Giannini for calling him a liar about Byrd's flight to the North Pole in 1947. I begin to suspect there was something mighty fishy about why the Department of the Navy in Washington refused comment on whether Admiral Byrd made a flight to the North Pole in 1947. I had written a letter dated April 18, 1960, to Admiral Arleigh Burke, Chief of Naval Operations,[1] about

---

[1] Arleigh Albert Burke (1901–1996) was an admiral of the United States Navy who distinguished himself during World War II and the Korean War, and served as Chief of Naval Operations during the Eisenhower and Kennedy administrations (1953–1963).

Giannini's claims only to find no reply to that letter. Perhaps this is why I wrote Giannini again without giving up.

Mr. Giannini has proven that there was an Arctic as well as an Antarctic expedition during the period 1946-1947, and now I know why the Navy refused comment. *[Editor's note: The following are quotes from Giannini's letter to Ogden.]* **Expressing the feature, little known, that there was an Arctic & Antarctic expedition during the period 1946-1947.** There follows excerpts from *The New York Times'* accounts during the month of February 1947:

Feb. 6, 1947: "Present Antarctic Maps Become Waste Paper."

Feb. 8, 1947: "Izvestia Sees Threat In U.S. *Arctic* Moves."

Feb. 9, 1947: "Pronouncement by Rear Adm. Cruzen, 2nd in Command of U.S. Antarctic Expedition:[2] "I have been in the polar zones for extended periods on three occasions. I am first to admit that I, too, am only beginning to understand the full facts about either pole."

Feb. 23, 1947: "(Frigid, Alaska) Frigid Force on Trek." (Arctic)

Feb. 24, 1947: "Byrd Quitting Antarctic Base, Ready to Leave."

Feb. 24, 1947: "Army Fliers Found with Downed B-29 Safe in Arctic." "The plane was one of a number which have been making routine photographic and mapping trips in the Arctic for many weeks."

Feb. 25, 1947: "FOR THE FIRST TIME, THE PURPOSE OF THESE FLIGHTS (ARCTIC), HAS BEEN GUARDEDLY ADMITTED."

Feb. 26, 1947: "The Navy announced that the aircraft carrier, *Philippine Sea*, is due at Quonset, R. I. tomorrow from Antarctica."

Feb. 28, 1947: "Rear Admiral Cruzen, (at Antarctic) speaks to press."

---

[2] Richard Harold Cruzen (1897–1970) was a decorated United States Navy officer with the rank of Vice Admiral best known for his participation and leadership in Antarctic expeditions.

ch. ends p. 358

Note: Contradicting instance of recent communication the foregoing is to be found on microfilm ending period designated.

Was it essential that Byrd wait for the *Philippine Sea*[3] to arrive at Quonset?[4] Could he not have flown to the Arctic base on Feb. 22nd?

And when did we desire that the Soviets be informed of our Arctic moves? Observe the above reference of Feb. 25th.

Whatever may be discovered, this author is pleased that Mr. Palmer has something to play with. The play may be instrumental in selling a few more books. The present book edition entailed an expenditure of $3,000.

No, Giannini did not make Giannini a liar. Nor did or could Palmer and all of this nation's confessed theorists make Giannini a liar.

In the event copies are required, this author will autograph and personally color the illustrations on all books ordered directly from this author.

And in closing, this author notes that Palmer and associates have made no effort to contact the Office of U. S. Naval Intelligence at the Pentagon. Why?

Since the Feb. 25th reference speaks of "guardedly admitted" who other than Naval Intelligence would possess data concerning the reason for the guardedness?

How does Palmer truly desire the facts, or is he only interested in controversy for the questionable uplift of his reading audience? It should be recalled that this author originally referred Palmer to Naval Intelligence as well as the *Times*.

Sincerely,

Giannini

---

[3] Launched on September 5, 1945, the USS *Philippine Sea* (CV/CVA/CVS-47, AVT-11) was one of 24 Essex-class aircraft carriers of the United States Navy, and the first ship to be named for the Battle of the Philippine Sea. It served in Antarctica as a part of Operation Highjump, 1946–1947.

[4] Formerly the location of Naval Air Station Quonset Point, a United States Naval Base in Quonset Point, Rhode Island, that was deactivated in 1974.

P.S. This author wrote: Byrd flew 1,700 miles beyond North Pole, not South Pole in 1947. And this author's book was unconcerned with the 1947 South Pole activity. But it did describe the South Pole activity of January 1956.

●

Mr. Giannini asks: "Was it essential that Byrd wait for the *Philippine Sea* to arrive at Quonset? Could he not have flown to the Arctic on February 22nd?" A very good question and the Navy refuses comment. There appears to be nothing ruling out the possibility of a secret flight by Admiral Byrd after February 22nd, 1,700 miles beyond the North Pole. Newspaper accounts show mighty strange things were going on in the Arctic.

Mr. Giannini speculates that Admiral Byrd left the Antarctic on February 22nd and made a secret flight to the Arctic. According to *The New York Times* of April 15, 1947, Admiral Byrd arrived in Washington on April 14th aboard the *Mount Olympus*,[5] flagship from the Antarctic and docked at the Washington Navy Yard.[6] Therefore, it doesn't seem possible that Byrd made a secret flight to the North Pole; yet the fact remains that Admiral Burke, chief of Naval Operations, refused to answer my letter of April 13th asking him if Byrd made that secret flight. If he didn't, why didn't the Navy answer. In view of this I don't think Giannini can be dismissed. Navy appears to be covering up for some sort of military secret or my letter would have been answered. What that secret is I don't know. But I do know one thing:

[5] Launched October 3, 1943, the USS *Mount Olympus* (AGC-8) was a Mount McKinley-class amphibious force command ship, named for the highest peak in the Olympic Mountains of the State of Washington, and was decommissioned on April 4, 1956.

[6] Established 1799, making it the oldest shore establishment and base of the United States Navy, the Washington Navy Yard (WNY) is a ceremonial and administrative center located in Washington, D.C.

ch. ends p. 358

The Admiral is [*censored -Editor*] for not answering that letter. So don't tell your readers that I dismiss your saucer theory. I can't disprove it any more than anyone can disprove Giannini. The whole thing is a bottomless mystery, thanks to the Navy.

I have a letter dated May 16, 1960, from Major Lawrence J. Tacker, USAF,[7] Public Information Division, Office of Information, regarding your saucer theory. He says: "The Air Force has no knowledge of the discovery of an oasis near the South Pole."

I believe you have an obligation to your readers to stop switching your arguments about Byrd from the North Pole to the South Pole but to go back to what you originally stated in your article "Saucers From Earth." Mr. Giannini deserves credit, not discredit.

Sincerely,

Richard Ogden

●

P.O. Box 695
New Providence, New Jersey
July 30, 1960
Dear Mr. Ogden:

Supplementing the hurried reply of July 27th, please be informed as follows: This author was extended courtesy, by the New York office of U.S. Naval Research, to transmit a radio message of godspeed to Rear Adm. Richard Evelyn Byrd, U.S.N., at his *Arctic base in February* 1947.

At that time, the late Rear Admiral Byrd announced through the press, "I'd like to see land beyond the pole; that area *beyond the Pole* is *the center of the great unknown.*"

[7] Lawrence James Tacker (1917–1996) was a Lieutenant Colonel in the United States Air Force, who authored the 1960 book *Flying Saucers and the U.S. Air Force: The Official Air Force Story.*

Subsequently Admiral Byrd, and a naval task force, executed a seven hour flight, of 1,700 miles, over land extending beyond the theorized North Pole "end" of the Earth. In January 1947, and prior to the flight, this author was enabled to sell a series of newspaper features to an international feature syndicate only because of this *author's assurance to the syndicate director that Byrd would in fact go beyond the imaginary North Pole point.*

As a result of this author's prior knowledge of the then commonly unknown land extending beyond both pole points, and after the syndicated features had been released in the press, this author was widely investigated by the Office of U. S. Naval Intelligence. That Intelligence investigation stemmed from the fact of Byrd's definite confirmation of this author's revolutionizing disclosures. (The book records that feature of investigation.)

Later, March 1958, this author delivered a radio address in Missouri, expressing the import of land discovered beyond the imaginary *North Pole* and South Pole points of archaic theory. Interest in that radio address was eloquently expressed by radio listeners in Missouri, Kansas, Iowa, Nebraska, and Oklahoma. (Their letters, about 500, are at hand at this date.)

And one of the most interested listeners was the local secretary of the U.F.O. And, if I am not mistaken, your able Ray Palmer's magazine is conducted in the interest of the U.F.O. organization.

The secretary requested that this author personally address the local U.F.O. chapter; and we willingly agreed. But it developed that an appropriate meeting place or auditorium could not be acquired by the local chapter due to lack of funds.

Prior to that radio address of 1958, and after the extensive land discovery *beyond the South Pole* point in January 1956, the U.S. Chief of Naval Operations, Admiral Arleigh A. Burke,

ch. ends p. 358

U.S.N., personally directed this author's basic treatise for adequate study by The Technical Assistant for Polar Projects in the Office of the Deputy Chief of Naval Operations for Fleet Operations and *Readiness*, at The Pentagon. (The Admiral's letter to this author is at hand; and photostatic copy may be had.)

Simultaneously, request for this author's basic treatise was received from diplomatic agencies of all other nations holding land bases beyond the South Pole point—the assumptive "end" of the Earth, south—.

Now, during the past three months, this author was invited to luncheon by a member of the late Rear Admiral Byrd's family. And that gentleman, who had read *Worlds Beyond the Poles*, observed: "Your book has changed the Byrd family's outlook concerning Admiral Byrd's accomplishments and elevated their personal esteem for him."

Significantly, it was not until then that this author was afforded realization of the Byrd family's previous ridicule of Admiral Byrd's honest appraisal of land he discovered *extending beyond and out of bounds of the* Copernican Theory's assumed Earth "ends," both *north and south.*

This author has graciously afforded you and Ray Palmer considerable wordage at no little effort. And this author is not subsidized by any agency, scientific or otherwise.

When this author answered your original communication, he reasonably assumed that *The New York Times* would have quoted Byrd's statement and the other press dispatches of February 1947. Further, this author then had no intention of a prolonged discussion with you or Ray Palmer. But, in consideration of the interest expressed by Palmer and your good self, we are convinced that you and the reading audience served by Palmer should be in possession of all the facts.

However, we are impelled to take a very dim view of Palmer's facetiousness in such forlorn expressions, as you have quoted: "We would have wound up talking to ourselves."

That manner can hardly compliment Palmer's intelligence. For we have at hand most commendable expressions relating to comprehensions of the book content from the following, to mention only the more important:

- Director, National Science Foundation,
  Washington, D.C.
- Director, The Arctic Institute of North America,
  Montreal, P.Q.
- Director, Adult Education, St. Louis University,
  St. Louis, Mo.
- Librarian, Pope Pius XII Memorial Library,
  St. Louis, Mo.
- Librarian, Boston Public Library, Boston, Mass.
- Librarian, Cambridge Public Library, Cambridge, Mass.
- Librarian, Baker Library, Dartmouth College,
  Hanover, N. H.
- Librarian, College of The Holy Cross, Worcester, Mass.
- Librarian, Loyola University, Los Angeles, Cal.
- Librarian, Notre Dame University, Notre Dame, Indiana.
- Capt. W. I. Martin, U.S.N., Commander,
  Barrier Force, U.S. Atlantic Fleet
- Rear Admiral G. L. Russell, U.S.N., Commandant,
  Twelfth Naval District, San Francisco, Cal.
- Commander Thomas R. Weschler, U.S.N.,
  Personal Aide to The Chief of Naval Operations.
- The Consul General of Spain
- The Consul General of Finland
- The Consul General of Denmark
- The Consul General of Belgium
- The Consul General of New Zealand
- The Consul General of Chile
- The Consul General of Mexico
- The Consul General of The Union of South Africa
- The First Secretary, Australian Embassy

ch. ends p. 358

- The Chairman, Committee on Science and Astronautics,
  U. S. House of Representatives, Washington, D.C.
- His Eminence, Francis Cardinal Spellman,
  Archbishop of New York, N.Y.C.

The foregoing have in fact received this author's treatise and/or book. And their intelligent and commendable expressions to this author fail to indicate that the book features presented developed the least amount of delirium or caused them to "talk to themselves." And should you or Palmer doubt the authenticity of the foregoing list we shall be pleased to send you or Palmer, or both, photostatic copies upon payment of reproduction costs plus postage.

Your interests, and the best interests of Palmer, can be best served by an intelligent and dignified application to the transcendent values described in this author's original work, *Worlds Beyond the Poles and Physical Continuity of the Universe*.

And Ray Palmer may be sure that "**Giannini did something in writing that book.**" He did too much after thirty-three years of application without subsidy from any source. (And, if you doubt the thirty-three years, there has been preserved press copies which describe this author's premature disclosures to the late Capt. Sir Georg'e Hubert Wilkins and the late Rear Admiral Richard Evelyn Byrd in *August* 1928.)

Mr. Palmer is considerably at fault in his quoted contention, "Giannini stakes his verity on *The New York Times* accounts." Giannini does no such thing. This author simply replied to you, as an unknown individual, with a conclusion assumed to be so and without any interest as to whether *The New York Times* had or had not carried the dispatch of other New York publications of that particular time. We depend on no publication; for we are cognizant that no publication holds all the facts. Further, the legitimate book account *does not mention The New York Times.*

Those accounts described Byrd's 1,700 mile flight, of seven hours, over land and fresh water Lakes **beyond** the assumptive North Pole "end" of the Earth. And the dispatches were intensified until a strict censorship was imposed from Washington.

As concerns your own too hasty observations, be informed that this author made no "blunder" of fact, nor did he in any manner "destroy" his book," as you so wrongfully assert. Neither you or Palmer, or his combined reading audience, can destroy the book or in the least measure detract from its transcendent features.

Neither you, or Palmer, or his reading audience, **know** where Rear Admiral Richard Evelyn Byrd was during the *entire month of February* 1947. But the conservative and revealing *Times* accounts recently transmitted to you, (and which you did not want to see) should afford a more intelligent appraisal for your own benefit.

Concerning your issue with Palmer's contention that Byrd flew 1,700 miles beyond the South Pole, you should now have realization that the 1,700 miles applied to the February 1947 flight beyond the North Pole. But, and this little morsel you can believe or reject, as you will, Rear Admiral George Dufek,[8] acting under Admiral Byrd accomplished a flight of 2,300 miles *beyond the South Pole on January* 13, 1956. Does that little feature disturb? It fails to disturb members of Naval Intelligence with whom this author has spoken since the occasion.

Would you, or Palmer, like to have the U.S. Chief of Naval Operations or the magnificent White House hero of the

[8] George John Dufek (1903–1977) was a Rear Admiral in the United States Navy, naval aviator, and polar expert who spent much of his career in the 1940s and 1950s in the Antarctic, first with Admiral Byrd and later as supervisor of U.S. programs in the South Polar regions.

ch. ends p. 358

American mob kiss you tenderly on the cheek and affirm, "Yes, Ogden, that is correct"? This author's studied conclusion is: don't wait for them to do so.

Ogden, in all seriousness, this author commits the following to you and to Ray Palmer: You abide on the west coast where it should be possible to develop film industry interest in a cinema dramatization of *Worlds Beyond the Poles*. Your reward, and Palmer's reward, for any material contribution making for such development would be most magnificent, indeed. And, since you claim that you do like facts, you may be assured that such widespread dissemination of book features would at long last bring out all the facts. For it would then be too late for the frightened rabbits, in Washington, D.C. and elsewhere, to attempt further withholding of the information which should have been made public property as early as the year 1947.

But do you really want the facts, or do you just want to make petty and empty controversy? For you have to date been guilty of gross misrepresentation in your reference to "fictional flight."

We do not know who you are or what your associations may be; but we do believe you are sincere. And, if we are correct in that conclusion, it would be well for you to *be sure of your facts*. *The New York Times* cannot provide all the facts, though we reasonably assumed it could until we spent six hours running a microfilm of 1947 *Times* editions. However, and as previously related, the Office of U.S. Naval Intelligence very definitely can provide the confirmation required; the U.S. Naval Research Bureau can do likewise; and the outgoing White House hero can do the same. **But**, but, but, will they. That would require a measure of moral stamina which is almost too much to expect.

The so-called "peasant" Soviet government has long been in *possession of all the facts* expressed most articulately by this author's work. And that possession of facts by the

Soviets, plus the *Soviets most able exploitation of such facts*, represents the paramount reason why this great nation has been compelled to suffer the gross indignities heaped upon it, and its representatives, in a manner heretofore never tolerated from any government or combination of governments.

This author cares not a little how you, or Palmer, or Palmer's magazine, use the name Giannini. But inasmuch as the name has been known and respected for more than 60 years, from Vancouver to San Diego—due to the activity of the late Amadeo Peter Giannini of Bank of Italy and Bank of America fame—Ray Palmer will not be contributing to the growth of his publication on the west coast, at least, if he continues to be indiscreet in his presentation of that name Giannini.

On the other hand, and as recorded, there is considerable to be gained by an *honest and truthful* evaluation of this Giannini's transcendent book features. Therefore, we shall be pleased to hear from you and/or Palmer whenever you and Palmer are pleased to express serious reasoning compatible with the sensational discoveries and unprecedented application of our time.

Sincerely,

Giannini

c/ W.T.P.

H.M.B.

F.A.G.

❧

Before we make pertinent comments on Mr. Giannini's letter, we wish to assure him we are not "playing," nor are we trying to uplift our reader audience (although what editor doesn't welcome new readers!). We are personally vitally interested in the truth, and if Mr. Giannini's book is truth, we are vitally interested in his book. Much has been said concerning the fact that we have suggested the Earth may

ch. ends p. 358

be shaped like a "doughnut." Let's clarify that right now: there is only one "fact" established to date, and only one "statement of theory" made by us, and that is simply that the Polar Areas are not as simple as our present maps show, but something **mighty peculiar** is true about both! And we intend to find out what it is. Our personal opinion remains unchanged, the Earth is a sphere, but that its **surface** is not necessarily spherical although it **appears** to be. Those appearances are both optical and astronomical. We have demonstrated quite conclusively (and more information will certainly be forthcoming) that the "horizon" is an optical illusion common to both lenses and the human eye, and that the eye or the camera are no "proof" that the world is round, and that to offer the "hull-down" phenomenon as **proof** of sphericity is totally irrelevant. We conceive of the possibility that an Earth, viewed from Mars for example, with a hole at the Poles would appear to be a perfect sphere with either ice-caps or cloud cover at the Poles. If there is a hole in Mars at the Pole, it could be quite true that it is concealed by a cloud cover which is erroneously called an "ice-cap." Mercury, during the recent transit across the face of the sun, developed a bright spot on its diameter which was the same intensity of the sun's face. Astronomers theorized that the bright spot was **on** Mercury, whereas your editor theorized it was the sun's brightness shining through a Mercurian "hole." **Both**, of course, are **theory!**

Next, we made no effort to contact Naval Intelligence for two reasons:

1) Mr. Ogden had done it. Our readers do much of our "leg work," thanks be!

2) Our experience with Naval (and other) Intelligence has been quite bitter.

Let us say that we consider any contact with them on the part of a civilian as "wasted effort," and all that can come of it is a lovely "run-around," if not a very real "push-around."

We **have** been pushed around by the military Intelligence agents, not only mentally, but physically. It is an experience we're not likely to forget, and one we won't let **them** forget. When a citizen of this country can be arrested without warrant, confined, denied access to legal counsel, or to communicate with friends or family, and can be knocked flat on his back, insulted, lied to, framed, coerced, threatened, and then finally, released and told to "keep your nose clean," Ray Palmer is the **wrong** citizen to try it on. If the military has a legitimate secret it wants kept, all this citizen asks is a courteous request to avoid spilling any vital beans. Punching around is not the sort of a request we can comply with. It is a fact that the Air Force has publicly announced that "flying saucers do not exist," therefore this editor believes that there is no military secret to be kept, and considers anything relating to flying saucer research (and this "polar area" as a possible origination point for flying saucers is related) is a perfectly legitimate field for civilian inquiring. As a civilian, we reserve the right to chase "phantoms," if phantoms they be!

Now, concerning Giannini. All we know to date is that Byrd **could** have made a North Pole flight in February 1947. Giannini is 100% right there. It does not mean that he did, however.

Aside to Mr. Ogden: If Major Tacker says the Air Force has no knowledge of an oasis near the South Pole, we refer him to the "Bunger Oasis"[9] and suggest that as Chief of the Public Information Division, Office of Information, he acquaint himself sufficiently to perform the duties of his office. If he is truly so lacking in information, we suggest to his superiors

---

[9] Bunger Hills, also known as Bunger Lakes or Bunger Oasis, is a coastal range on the Knox Coast in Wilkes Land in Antarctica, consisting of a group of moderately low, rounded coastal hills, and notably ice free throughout the year, lying south of the Highjump Archipelago. The reasoning behind the minute amount of ice in the area is still relatively unknown and remains under intense debate amongst scientists today.

ch. ends p. 358

that he might be profitably replaced by a more informed man. His salary definitely comes under the heading of "governmental waste." As for my "switching around from North Pole to South Pole," we're not—we're concerned with **both**, but we're hanged if we know which is which in our information, particularly in the case of Giannini. Until we have it straightened out, there is bound to be a lot of confusion, even in this editor's mind! How else can we straighten it out if we don't present all the pros and cons?

Mr. Giannini: if the New York office of U. S. Naval Research extended you the courtesy of transmitting a radio message of Godspeed to Byrd at his **Arctic** base in February 1947, and that message was transmitted, can you condemn us for suggesting that perhaps you have the documentary evidence to support this? If you have, there is no question as to a Byrd North Pole flight in February 1947.

How could you assure the "syndicate director" that Byrd would actually fly over the North Pole beyond that imaginary point without actual documents? May we have such evidence?

Your basic treatise, directed by Admiral Burke to the Technical Assistant for Polar Projects could not have been done without documentary action. Do you have such documents? You say you do, and even if so, why would it be evidence of the truth of your treatise? In that respect we refer to the long list of people who "did not talk to themselves" when informed of your theory. In your book, **[on page 17]**, you tell of a meeting with Cardinal William O'Connell wherein you outlined your theory to him. His comment was: "If it is so, the world will know of it." So far as we know, the reactions of each one of those persons you list who gave you "commendable expressions" was precisely on the order of the O'Connell comment. This editor submits that this is the polite comment, expressing nothing positive, but possibly even concealing a quite negative reaction. O'Connell is not at all convinced. He says "if it is so," it will all come out in the wash. In effect, he is saying "but **you** haven't proved it to be so at all!" He's just being polite to another of those visitors we all have who take up our time with matters we're really not interested in, or in which we do not concur in the least. Rather than brusquely saying "I don't believe it is so," or worse, "bunk!", we say something polite.

If an author (you, as example) send copies of your treatise to a hundred prominent people, you should get a hundred polite expressions, none of them condemning the treatise. To say that they are "evidence" for the acceptance of the theory is wishful thinking. We wonder how many publishers said to you: "Your book is extremely interesting, but not quite in our line." That is probably why you had to spend $3,000.00 to get it printed yourself. Not that a publisher's opinion not to publish means a thing, mind you! Many publishers have turned down truly great books, to their chagrin later on. But likewise, their polite comment does not constitute an endorsement.

Requests from diplomatic agencies of all other nations holding land bases "beyond the South Pole point" is **certainly** evidential! We would appreciate such documentary evidence, and offer you our pages to present it to back up your book! You should have used it in your book — or is it all "confidential" and you are instructed not to reveal the fact that the request has been made? But you say to us that all the other nations requested it, so the secret is out. Best let us back it up with the document itself.

As for you being "subsidized by any agency, scientific or otherwise," we are going to take the opportunity to enlarge upon our own thought in this respect. If you mean anybody described as a private citizen or group of citizens, or a private scientific or business firm has not subsidized you, we have no argument. We accept your word. The "otherwise" bothers us. If the "otherwise" was a governmental agency whose

ch. ends next p.

purpose was confusion and cover up of the flying saucer story, then of course the "otherwise" becomes meaningless, because such subsidy would automatically also carry with it the connotation of deceit — you could not be subsidized by such an agency, and at the same time admit it! So, the "otherwise" is extraneous to your assertion, and should properly have been left out. Major Tacker, as an example, would have left it out. His training would have dictated it.

To close our comment, let's suggest to our readers that in this editor's mind, it is **quite possible** that not only is Giannini "subsidized," but also Mr. Ogden. Presumably Mr. Ogden is a reader of *Flying Saucers* who is enormously interested in flying saucers, and determined to furnish us with every bit of evidence he can. We appreciate that, and it makes no difference to us if he is "subsidized" or not. Either way, his help is valuable. As O'Connell put it to you: "If it is so . . ." it will come out in the end, and your editor is a patient man.

Mr. Ogden, don't go off into the wild blue yonder because of the foregoing — I am merely stating a possibility. In the sort of investigation your editor is engaged in, the scope of "possibilities" must always be included in all thinking. For example, I would probably not be in flying saucer research today at all, except for a "possibility"; a "beauty parlor operator" whose unlisted phone number could be discovered by asking for it at Marine Intelligence at the Federal Building!

Byrd Revises Map of Antarctica; Fliers Find Lofty Peaks, Vast Sea

Special to THE NEW

LITTLE AMERICA, Feb. 5—
Rear Admiral Richard E. Byrd revealed today details of the Navy's recent discoveries that turn the present maps of Antarctica into waste paper.

I have been in the Polar zones for extended periods on three major research, scientific and development expeditions. I am the first to admit I, too, am only beginning to understand the full facts about either Pole, for no man in one lifetime is able to acquire the entire complex knowledge to make him a positive authority.

But many men have come way towar

IZVESTIA SEES THREAT IN U. S. ARCTIC MOVES

Special to THE NEW YORK TIMES.

MOSCOW, Feb. 7—The United States military activity in the Arctic threatens the interests of peace, according to the viewpoint advanced in an article in the Government newspaper Izvestia.

The Measures taken by the United and Navy

ES, SUNDAY, FEBRUARY 23, 1947.

FRIGID FORCE ON TREK

14 in Alaska Unit to Travel 150 Miles in Five 'Weasels'

HEADQUARTERS, Task Force Frigid, Alaska, Feb. 17 (P) (Delayed)—With daily air contact planned arrangements for med

NEW YORK, MONL

BYRD QUITTING BASE FAST TO ESCAPE ICE

Trip North From Little America Rushed — Western Group's Planes Extend Discoveries

The Navy's Operation Highjump in sts.

For the first time the purpose of these flights has been guardedly admitted. Officers, quoted by the United Press, assert that it is to "prepare against possible invasion from the Eurasian land mass." Land, reported that the weather was good and that it hoped to launch an exploratory plane.

WASHINGTON, Feb. 26 (U.P)—The Navy announced today that the aircraft carrier Philippine Sea is due at Quonset, R. I., tomorrow on her return from the Antarctic.

ARMY FLIERS FOUND WITH DOWNED B-29 SAFE IN THE ARCTIC

Search Craft From Alaska Drop Food to Eleven Men Missing for Two Days

WATER OPEN FOR RESCUE

eaplane Can Pick Up Crew, hose Radio Told of Plight in the Area of Greenland

Light of Illusion

Light that's seemingly so far,
You are not a detached "star";
And no mystery can be,
Of your shining quality.

Though your "twinkle" seems to be,
It's a trick eyes play on me;
For I've learned how they deceive,
And illusory image leave.

As patch of outer celestial sky,
You're bewitching to the eye;
Yet you cover unseen land,
As does earthly sky at hand.

You know not isolation's plight,
Though presenting lonely sight;
For you're linked in sky embrace,
Common to this earthly place.

And at last I'm on my way
To visit 'neath your bright display;
I won't have to move through space
In fantastic rocket pace.

Straight ahead from polar region,
over land and waters legion,
Moving in established manner,
I'll reach your celestial manor.

—F. AMADEO GIANNINI

Francis P. Giannini

HONORARY HEATHEN

Admiral Byrd Calls Navy Survey Of Antarctica Notable Advance In Geographic Study Of The Earth

Washington, D. C.—Rear Admiral Richard E. Byrd, writing in the National Geographic Magazine, expressed the opinion that the Navy's 1946-47 Antarctic Expedition, which he headed, brought a notable advance in man's knowledge of the earth.

"This expedition accomplished more in the way of increasing general geographic knowledge of the south polar regions than any other expedition," the noted explorer said, adding parenthetically: "This does not apply, of course, to detailed scientific research.

"Together with the three previous expeditions I have led, more was accomplished in geographic discovery than by all other Antarctic expeditions combined."

Titanic Laboratory

Describing the Antarctic as a "titanic physical, chemical and biological laboratory," Admiral Byrd summarized accomplishments of the 4,000-man expedition:

"During the flights of our three groups an area more than half as large as the United States was covered. Of this, at least 340,000 square miles never had been seen by man before. It also was possible to explore about 75,000 square miles of ice-strewn ocean where no ship had ever sailed.

"More than 5,400 miles of coastline were discovered, relocated, or confirmed. Counting bays and indentations, the total would be considerably greater.

"Ten new mountain ranges, among them some of the loftiest on earth, were discovered. New archipelagoes, peninsulas, islands and seas were placed on the map. Some of the world's largest glaciers were found and photographed. Great extensions were made in the known area of the enormous Antarctic plateau, approximately 8,000 to nearly 11,000 feet above sea level, and for the first time part of the vast plateau beyond the South Pole from the direction of the Ross Sea was explored.

"There were unexpected discoveries, such as that of the large ice-free region of open-water lakes near the edge of the continent.

"Out of the whole emerges a new and more accurate picture of the land, which is like a titanic upturned bowl just below the stratosphere with a badly cracked rim of mountain ranges."

Huge Coal Reserves

Admiral Byrd, first man to fly over the North and South Poles, said that eventually someone may make money out of the Antarctic continent, nearly as large as South America.

"We know, for example, that there are huge reserves of coal there," he wrote. "The black mountains are full of it. It is impossible at the present stage of exploration even to make a wild guess as to the extent of these deposits.

"But any mining operations, especially when we consider the difficulties of transport, would be fantastically impractical at this time. There may come a day, however, when the world will need this coal.

"Almost certainly oil will be found under the ice. It is impossible to imagine a large continent without vast mineral wealth of many kinds buried in its rocks.

"I dislike to think of money in connection with Antarctica. It has higher values. This continent and these seas can be looked upon as Nature's most sublime work of art. They are poetry, music, painting, architecture, and philosophy all combined."